Crossing the Hudson

Crossing the Hudson

Historic Bridges and Tunnels of the River

DONALD E. WOLF

Rivergate Books

AN IMPRINT OF RUTGERS UNIVERSITY PRESS
NEW BRUNSWICK, NEW JERSEY, AND LONDON

Library of Congress Cataloging-in-Publication Data

Wolf, Donald E., 1927–
Crossing the Hudson : historic bridges and tunnels of the river / Donald E. Wolf.
p. cm.
Includes bibliographical references and index.
ISBN 978-0-8135-4708-4 (hardcover : alk. paper)
1. Bridges—Hudson River (N.Y. and N.J.)—History. 2. Tunnels—Hudson River (N.Y. and N.J.)—History. I. Title.
TG25.H83W65 2010
624.209747′3—dc22
2009016198

A British Cataloging-in-Publication record for this book is available from the British Library.

Visit our Web site: http://rutgerspress.rutgers.edu

Manufactured in the United States of America

Frontispiece: Erecting the 500-foot-long, 1,800-ton truss assembly that would complete the superstructure framing for the second Newburgh-Beacon Bridge in August 1980. Photograph reproduced by permission of the American Bridge Company, which fabricated and erected the steel for the bridge.

CONTENTS

ILLUSTRATIONS

ACKNOWLEDGMENTS

In writing *Crossing the Hudson*, I have not ignored the technical elegance or the physical beauty of the bridges and tunnels that cross the river, as they are clearly the very essence of these wonderful structures, but I've left the detailed exposition and analysis of those qualities to others. My objective here is simply to provide a history for the reader who wants to know where, why, when, and how these structures were built, something about the people who designed and built them, and perhaps a little about the social, political, and technical problems that had to be solved before, while, and sometimes after the structures were built.

Of course, few resources could be more vital to the process of ferreting out all that information than the libraries, museums, and historical societies that ensure the preservation and availability of the relevant manuscripts, books, and other materials, and I'm well into the debt of many of those institutions and of the good people who make them work.

In the Hudson Valley alone these places include the New York State Library at Albany, a superbly managed institution whose staff and collections are hard to match anywhere. At the south end of the river, the New York City Public Library System, most especially its Science, Industry, and Business Library (SIBL) division, was invaluable, largely because of the special efforts of Marguerite A. Nealon, whose title there is Science Bibliographer. But not all the libraries that contributed so effectively were in the Hudson Valley. The Linda Hall Library in Kansas City, Missouri, has a vast collection of historically important engineering and scientific material that was central to my research into the histories of these bridges and tunnels, and the energetic and always thorough help of Christine Taft at that library was what made it all work for me.

The Hudson Valley is lined with other libraries and historical societies that contributed to various parts of the story. In Waterford, where Brad Utter is the capable and amiable director of the Waterford Museum and Cultural Center, I found an abundance of material about the river's first bridge. Just downstream, in Troy, Elsa Prigozy, a volunteer at the Rensselaer County Historical Society, made it easy to secure what I needed to know about the Green Island Bridge (the river's first railroad bridge); and the nearby Folsom Library at the Rensselaer Polytechnic Institute (RPI) was another valued resource in Troy. The Vedder Library in Coxsackie, the Hudson Public Library, the Catskill Public Library, the Kingston Public Library, the Cornwall Public Library, and the Adriance Memorial Library in Poughkeepsie were all first-class sources, especially for material about the four bridges that cross their long section of the river. In Newburgh, Rita Forrester in the Local History Room of the Newburgh Free Library was tireless in making available to me an abundance of helpful material about the Newburgh-Beacon bridges. Michelle Figliomeni at the Orange County Historical Society was able to provide material about the Harriman family and its role in building the Bear Mountain Bridge that would not have been available anywhere else. In Peekskill, Tarrytown, and Nyack, the Field Library, the Warner Library, and the Nyack Public Library, respectively, were essential sources for probing the detailed histories of the Bear Mountain Bridge and the Tappan Zee Bridge. For the George Washington Bridge and for the tunnels that carry railroads and motor vehicles into and out of New York City, nothing can match the collections of the New York Public Library.

Beyond the consistently high-quality material made available by all those libraries and historical societies, I occasionally discovered a source of such extraordinary value that it deserves to be individually acknowledged. One of those was Jameson Doig's *Empire on the Hudson* (New York: Columbia University Press, 2001), a gracefully written account of the history of the Port Authority of New York and New Jersey. Doig's insights into and description of the growth and development of the Port Authority provided a context for understanding how this interstate agency designed and built its bridges and tunnels, and his book acquired a special value after most of the Port Authority's historical records were destroyed on September 11, 2001. The other book that proved to be equally valuable to me as an exceptional source was Henry Petroski's *Engineers of Dreams* (New York: Alfred A. Knopf, 1995). A clearly gifted writer and engineer, Petroski guides his reader

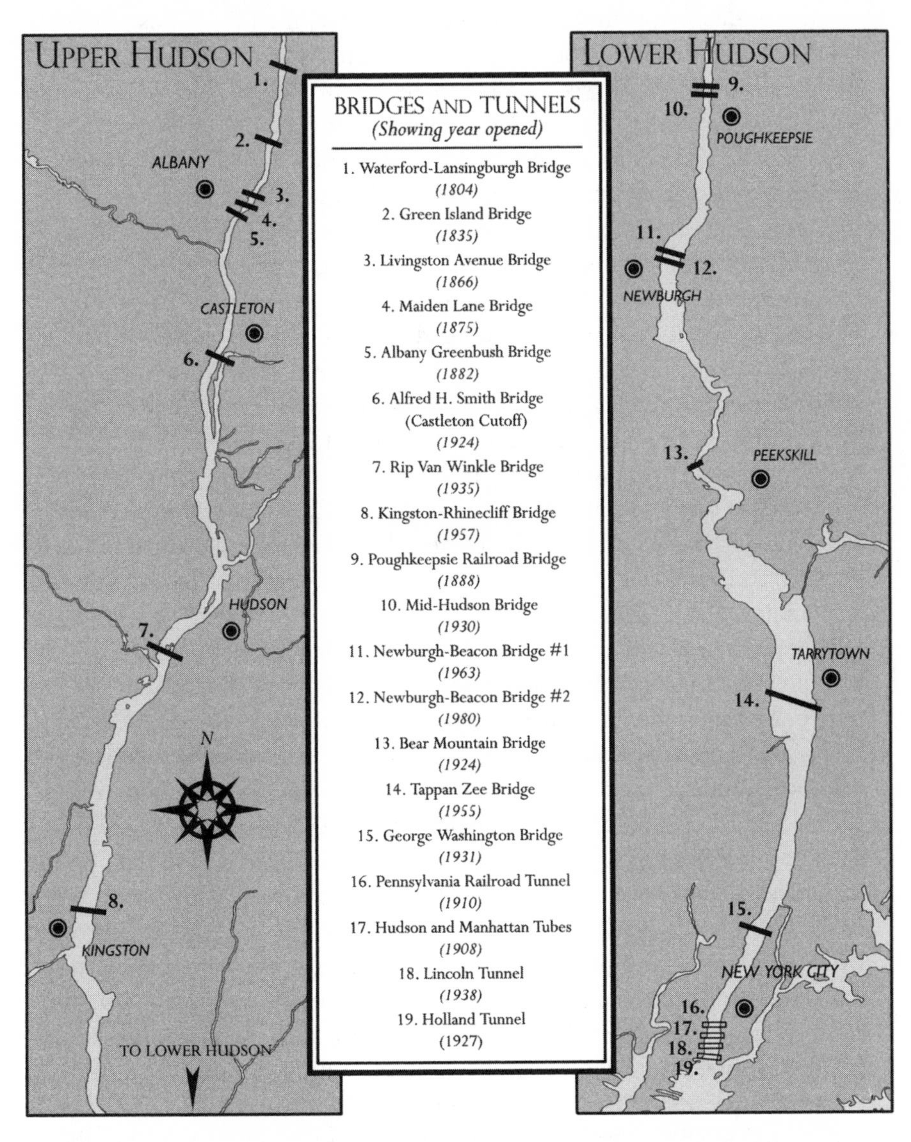

FIGURE 1 Historic Bridges and Tunnels of the Hudson, Showing Where They Cross (or Have Crossed) the River. Drawn by Zack Stella.

thirty years later at Castleton, near Albany. At the same time, American bridge builders had been developing ways of configuring and joining structural elements that would make it possible to lengthen the spans of new bridges, accommodating the complaints of ship captains who struggled to navigate their vessels between closely spaced supporting piers in the river. It wasn't long before the first graduates of the country's emerging engineering schools were adding the sophisticated mathematical and physical insights needed to make such longer spans safer. With concurrent improvements in iron and steel technology, it would soon become possible to carry even heavier loads without compromising the span lengths needed for shipping.

Early in the twentieth century, other technologies entered the picture. The McAdoo and Pennsylvania Railroad tunnels, relying on new tunneling techniques, allowed railroads to travel between New York and New Jersey below the river. Soon afterward, the internal combustion engine burst on the scene, producing complex needs of its own. Between about 1920 and about 1960 the Holland and Lincoln tunnels were designed and built to allow cars and trucks to cross quickly and safely under the river, too. The dramatic impact of motor-driven vehicles on American culture would between 1924 and 1940 induce the construction of three spectacular, long-span suspension bridges across the Hudson: the Bear Mountain near Peekskill, the George Washington at the northwestern edge of Manhattan Island, and the Mid-Hudson at Poughkeepsie. Just before World War II brought a temporary halt in the construction of new bridges, a long-span cantilevered bridge was built between Catskill and Hudson. After the end of the war, the volume of automobile traffic would grow at an even more furious pace, and its impact on state and federal policy would soon (to the dismay of the declining railroads) generate massive government subsidies for state and interstate highway systems. Along the Hudson, that meant major, long-span bridges at Tarrytown and Newburgh, where two of the new superhighways would cross. At about the same time, population shifts were combining with powerful social changes to place automobile travel at the center of the country's postwar culture, further intensifying the demand for bridges and leading to another one on the Hudson, between Kingston and Rhinebeck.

The chapters that follow tell the story of those two centuries of bridge and tunnel building on the Hudson. The process is continuing into the twenty-first century, and planning is already in place for at least one new rail tunnel between New York and New Jersey and for the complete replacement of the Tappan Zee Bridge at Tarrytown. Detailed exploration of those complex plans has been left to others.

through the professional lives and problems of some of the most important of the country's bridge designers, providing a wonderfully well-informed context for studying and understanding the work that was done on the Hudson.

Not every source for *Crossing the Hudson* is a book or an article. Some of the details came directly from especially knowledgeable informants, and they certainly deserve to be thanked here. One was Frank Griggs, an RPI-educated professional engineer from Rexford, New York, who has long experience in bridge design and is himself writing a book about the Waterford Bridge. Another was Richard Barrett, a historian whose knowledge of Albany, especially where it relates to the railroads of the region, is exhaustive. Barrett's contribution of information and of illustrations from his own collection is immensely appreciated. Still another person who provided valuable help was Dietrich Warner of the Rosendale Cement Company; his knowledge of the cements used in the early nineteenth century was useful and appreciated. The advice of Michael Cegelis, senior vice president of American Bridge Company, was tremendously helpful in researching the work done by his company on several of the river's bridges, and the recollections and research of engineer Ted Seeley were equally valuable. Seeley worked for the Snare Corporation when that company was doing foundation work for the first Newburgh-Beacon Bridge, and his father ran the company for a while, so his recollections had a special authenticity. Barney Martin's knowledge of the engineering on the Newburgh-Beacon bridges was uniquely helpful. He is now president of Modjeski and Masters, which designed them. One of the most informative persons I interviewed was Henry J. Stanton. A former executive director of the New York State Bridge Authority, he understands some of the most subtle issues in the history of these Hudson River bridges, and his help was critical to my understanding them.

Managing the design and construction of the upstate bridges has been mostly the work of two state agencies, the New York State Bridge Authority and the New York State Department of Transportation (formerly the New York State Department of Public Works), and both agencies were vital sources. The chief engineer at the Bridge Authority, Bill Moreau, was the moving force when it came to historical information, and John Bellucci, the agency's director of planning and public relations, was able to produce the needed pictures as soon as they were requested. At the Department of Transportation, George Christian, the deputy chief engineer, was never too busy to probe thoroughly for information that had long since been sent off to dead storage. Those documents were vital to the process and much appreciated.

In some cases, words are not enough to tell the stories of these bridges and tunnels, so illustrations are needed too. I'm in the debt of all the institutions and individuals who provided the photographs, and I've named them in the accompanying credits. My special gratitude goes to Zack Stella, an illustrator from Putnam Valley, New York, whose excellent map of the Hudson Valley elegantly clarifies the locations of all the bridges and tunnels described in the book.

So much for technical support. No married man could immerse himself in such a project as this one for as long as I have without the generous tolerance and understanding of a loving wife, and I'm no exception. My appreciation and love for her have only increased during almost sixty years of marriage.

Crossing the Hudson

Introduction

Bridges and tunnels are in some ways like great monuments. They have the same high visibility and inherent permanence that demand exhaustive public scrutiny and deliberation before they can be built, and like them, they develop histories of their own that say a good deal about why they were built, how they were built, where they were built, and about who built them.

Crossing the Hudson explores the histories of the historic bridges and tunnels that were built during the first two centuries that followed the end of the American Revolution. Before that war, not a single bridge or tunnel crossed the river, partly because the technology was not readily available, but mainly because a dozen or so ferries were already doing a decent job of satisfying the public's still modest need to get to the other side. The American Revolution made the importance of being able to cross a subject of serious interest. Although military reasons for ensuring reliable crossings declined after the end of the war, an array of social, political, industrial, and commercial factors emerged to take their place.

The issue of defending and expanding their ability to cross the Hudson first surfaced for Americans at the very beginning of the Revolution, when the British in New York decided to tighten their grip on the river as a way of preventing the rebellious colonists from transporting men and matériel across it. Their strategy was to split the colonies in two, separating them with a river the colonists would not be able to cross.[1] The Americans reacted aggressively by filling much of the lower Hudson with sunken ship hulls and then chaining it off in several places to keep ships from entering. None of these strategies worked. The British just sailed around or destroyed all the obstructions (with a little help from a few Loyalist sympathizers) and continued north toward Albany, where they planned to link up with Canadian troops under General John Burgoyne.[2] But the fortunes of war turned against them, and when they learned that an inspired American force at Saratoga had defeated and captured Burgoyne's entire army in October 1777, they turned their fleet around and headed back to New York, leaving control of the river to the Americans.[3]

A few years later, in 1780, the British tried again. This time they entered into a secret arrangement with Benedict Arnold, who—after years of loyal service under General George Washington—had been appointed commanding officer of the American fortification at West Point. The plan was for Arnold to turn over that strategically valuable military asset, but the plot was exposed before it had a chance to play out, Arnold's British accomplice was hanged, and Arnold, after briefly commanding British troops against American forces in New England and Virginia, went off to England. For almost twenty years after that, he drifted into and out of failing business enterprises there and in the maritime colony of New Brunswick, receiving occasional modest subsidies from the British government and ultimately dying impoverished in England at the age of sixty.[4]

Attention to river crossings waned for a few years after the end of the war, but as villages along the Hudson's banks began to flourish, the development of country roads and turnpikes served to identify towns where crossings were likely to make life better. Commerce and industry began to flourish, and by 1804 the first bridge across the navigable Hudson had been built between Lansingburgh and Waterford, just north of Albany and Troy. A hybrid timber arch and truss bridge on stone piers, it had four clear spans that averaged about 175 feet, and its novel design represented an important step forward in bridge technology. After Waterford, new bridges or tunnels along the rest of the river would come along at an average rate of about once every ten years for almost two centuries. Their planning, design, and construction are the subjects of the thirteen chapters of this book.

Technology was usually the driving force, but it was not always the technology of structural engineering that mattered most. In some of the earliest examples, builders were responding to the needs of a new transportation system, the steam-powered railroad. Pedestrians and horse-drawn vehicles had been crossing on ferryboats for years, but when steam railroads came along, there was no practical way to get them across the river without bridges. Between about 1835 and about 1890 five major bridges for carrying railroads were built at Troy, Albany, and Poughkeepsie, and a sixth would follow about

CHAPTER 1

Waterford, the First Bridge

EXCEPT FOR AN OCCASIONAL SPIRITED, sometimes bloody, but historically insignificant episode, the American Revolution in the Hudson Valley pretty much ended when Benedict Arnold failed in his attempt to hand over West Point to the enemy in 1780. In New York, the British under Sir Henry Clinton remained hunkered down with about 9,500 troops, waiting for a chance to attack the rebels in one place or another and only occasionally breaking out to do so, while George Washington's forces waited and watched from the surrounding hills. But not much really happened. The Hudson itself was still effectively blockaded by a huge fleet of British ships that had been anchored in New York Bay since 1776, so the pace of shipping up and down the river remained distressingly slow, and activity in the towns along its banks pretty much dried up. With most of the valley's young men still serving in the military, there was more than enough war weariness to go around, and it would still be some time before anyone would begin to think about bridges across the Hudson River.

Of course the war had not bogged down everywhere by 1780. Among the British there had been some sentiment for abandoning the campaign against the troublesome Yankees of the Northeast and settling instead for a less grand American empire that would include only the old British holdings in the West Indies and Canada and in some of the rebellious southern colonies, where the royals had been making some solid military gains. That view gathered enough support at the policy level to turn the British focus toward the American South that year, and by 1781 their forces had moved up through Georgia and the Carolinas and were positioned to do battle with the Americans at Yorktown, in Virginia. It was there that the Franco-American alliance that had taken root after the American victory at Saratoga was to prove its extraordinary value. With the help of a French fleet under the Comte de Grasse and a French army under the capable Comte de Rochambeau, Washington and his troops were able to defeat the British under General Lord Cornwallis. After that convincing American victory, what little remained of the war began to crumble. Although it would take a few years to extinguish a few lingering fires and to work out the details, the British withdrawal from American shores would be complete by the fall of 1783.[1]

Celebration of the long-awaited American independence would prove to be spirited but brief. It didn't take long for people to realize that seven years of war had depressed their economy and that they were in desperate need of a formal system for managing the new republic. During most of what was left of the 1780s, crime was a serious problem, counterfeit currency abounded, personal bankruptcies were common, and just about everyone except the farmers (who were profiting from the currency inflation) found themselves in bad straits. Bitterness against the former mother country would take a long time to wane, and it was exacerbated by what later proved to be an exaggerated report that more than eleven thousand American prisoners had died aboard British prison ships in New York Harbor. All the big Tory estates had been confiscated, and Loyalists who remained in the country continued to be despised and persecuted for a long time.

In spite of all that, the colonies did begin to recover. In New York State families were reunited, birth rates increased, farms were restored to productivity, and mills and shops began to function again. Life was slowly returning to normal. Despite deep antagonisms that still separated the mostly downstate advocates of strong central government from the mostly upstate advocates of virtually no government at all, a federal constitution that defined and favored central government would finally be ratified by the states in 1788, and in 1789 George Washington would be inaugurated as the country's first president.

Along the Hudson, which before the war had been crowded with vessels sailing from one river town to the next and even more often between river towns and the bustling seaport of New York, signs of increasing activity confirmed that prospects were improving. After seven years of British blockade, when commerce and travel along the river had slowed to a trickle, the Hudson Valley was more than ready to resume its growth. The vessel of choice was a graceful, approximately 75-foot-long, single-masted sailboat that carried a huge mainsail well forward, a relatively small jib, and a triangular topsail, had a capacity of about 100 tons and was called the North River sloop. Its name reflected the early practice of calling the Hudson River the "North River," to distinguish it from the Delaware, the "South River." Originally brought to America by the Dutch almost two centuries earlier, the North River sloop had almost disappeared from the region during the war years, but by the early 1800s more than a hundred of them could be seen carrying freight and passengers up and down the Hudson and their numbers were growing.[2]

That quickening of activity was one of many signs of the increasing postwar viability of the towns along the river's banks, and there were plenty of others. The original towns were expanding and new ones were appearing. In 1784 a small fleet of vessels arrived from Nantucket Island with eighteen families from a community of Quaker seamen and artisans whose loyalty to the colonies had been challenged by their neighbors when they had sought to continue their trade with the British during the war. Now they were seeking a more hospitable deep port inland, and they found it at Claverack Landing, south of Albany, where they would develop the thriving whaling center of Hudson.[3]

The river was again emerging as an important commercial artery. More families were settling in the region, and most of the families already established there remained and thrived. Less than two decades after the end of the war, the arrival of new settlers, many from the harsher environments of northern New England, would increase by almost 50 percent the population of what would eventually become the ten counties of the Hudson Valley. Commerce into and out of the river towns accelerated, with an increasing volume of goods passing through them on their way to or from inland locations.[4]

Crude paths and then wagon roads for transporting goods and people to and from the river towns evolved and were subsequently improved. Until the early years of the nineteenth century, responsibility for establishing and maintaining such roads had been assumed by the counties, which administered the process but passed along responsibility for actual construction to local overseers. The state's General Highway Law mandated that eligible males perform at least three days of road work each year or pay sixty-two and one-half cents per day to be excused. For the most part, such publicly supported efforts to build usable and durable roads were failures. Self-serving landowners from different areas impeded the work by competing with one another for ever-bigger slices of the public pie, and other landowners opposed tax expenditures for almost any purpose at all. Publicly supported road-building programs eventually disappeared and were succeeded by more effective programs of chartered and privately funded toll roads or—as they were usually called—turnpikes.

Some turnpike investors were local merchants who saw the proposed roads as likely to improve the accessibility of their shops, and some were landowners who felt that properly located, well-constructed roads would enhance the value of their land. Few investors saw shares in the turnpikes as likely to generate cash returns, and it was just as well they didn't. There wasn't much left after the "shunpikers"—who bypassed the toll collectors on back routes—had taken their free rides. Even without the impact of the shunpikers, there was never much left after the collectors had waived the fees of a long list of legally toll-exempt travelers that included anyone on the way to or from church, a voting place, the performance of military service or jury duty, a funeral, a grinding mill, a blacksmith's shop, or the person's private residence if it was within a mile of the toll house. Whatever remained in the till after all those exemptions had eroded the receipts was rarely enough to pay for maintaining the roads, and after about 1850 it wouldn't be enough to do even that.

But while the turnpike period lasted, a good many toll roads were built in New York State, substantially advancing the development of the area and almost invariably terminating at one of the increasingly important towns along the east and west banks of the Hudson. Growing populations and quickened commerce in some of those towns had by the early nineteenth century already begun to identify places that would one day be the sites of bridges.[5]

Such bridges, for the most part, were still nothing more than abstractions in the imaginations of people in the river towns, but that wasn't true everywhere. Across a narrow strip of the Connecticut River at Bellows Falls, Vermont, for example, Enoch Hale had in 1785 built a groundbreaking arch-truss bridge that connected two sections of a newly built turnpike between Albany and Boston. Drawings for the Hale bridge have disappeared, but surviving paintings show that its approximately 350-foot-long roadbed was carried across the river on four parallel stringer systems composed of spliced, built-up timbers 50 feet above the surface of the river and supported at mid-span by an elaborate arrangement of piers and span-shortening, triangular bracing. A remarkable structure for its time, the Bellows Falls Bridge lasted for forty-five years and fared better than its designer.

Hale had borrowed the money to build his historic bridge, and when a messenger failed to deliver his final payment on the day it was due, the lender exercised his right to assume ownership of the bridge. Having lost his bridge, Hale never designed another one and died in poverty a few years later.[6]

And there were other important early bridges, too. The country was still very young, and between its growing population and the vigorous westward migration that was starting to take shape, the clear promise that there would be increasing demand for roads and bridges ensured that there would be an abundance of persons ready and willing to design and build whatever was needed. Among those plucky enough to advocate and underwrite the cost of such early bridges was a small group of entrepreneurs in the towns of Lansingburgh (later part of Troy) and Waterford, which faced each other across about 800 feet of the Hudson at the extreme northerly end of its navigable waters. Around 1800, these fellows began talking about a toll bridge to connect their towns, clearly a risky venture in that still sparsely populated section of upstate New York, but apparently not too risky for men who saw the promise of growth in almost everything that was happening in the region. At that very narrow reach of river, a bridge promised to be short and relatively inexpensive, and because all river traffic of any real size had already been forced by the shallow depth of the water to remain south of the proposed bridge route, the normally expensive task of ensuring that river shipping wouldn't be obstructed by closely spaced piers didn't have to be faced. Certainly another compelling argument for a bridge was the substantial patronage enjoyed by the twenty stagecoaches that were already carrying passengers and freight between and among the river towns of Waterford, Lansingburgh, Troy, and Albany, mostly by ferries that were entirely at the mercy of the weather.

What might have been the most influential arguments of all for a bridge between Waterford and Lansingburgh were the new turnpikes that promised to be opened by about the time the new structure was expected to be completed. One of these, already substantially advanced, was to run between Lansingburgh and Granville, a New York town close to the Vermont border. At Granville there was a connection to a good road to Whitehall, and there was reason to believe that a spur would soon connect it with the town of Rupert, in Vermont. Altogether, that string of roads was the beginning of a viable connection between the bridge site and the fertile markets of New England. If connecting roads are the arteries needed to sustain the life of a toll bridge, the Waterford to Lansingburgh crossing was clearly in for a long life.

Understandably, the turnpikes seemed to the bridge planners to be a promising factor, but in the long run not all of them worked out as well as expected, and what ultimately became known as the Waterford and Whitehall Turnpike fared worse than most. After a few years it began to fall into disrepair, and its proprietors were indicted for perpetrating a public nuisance. Not long after that, an angry mob put an end to the turnpike itself by tearing down its last toll house. Business on the road had been declining, anyway, because part of the Rutland Railroad had been built close enough to its route to take away most of its remaining customers.[7]

Still, in 1800, the Waterford-Lansingburgh Bridge investors could see little but good times ahead, and they secured authorization from the state legislature to build their bridge. It would be the first bridge of any kind to cross the Hudson between the northern limit of the river's navigable waters and its mouth at New York Harbor. By 1803 the originators had established and sold shares in their newly formed Union Bridge Company and had begun a search for someone to design the bridge and build it.

That was no easy task in those days. It would still be a while before the university-educated civil engineer would come along, a person who had been trained to design bridges and other structures but who had in most cases never earned a living using carpenters' tools. The professional associations and directories that would in later years make finding someone to design a bridge easy had not yet appeared. Few Americans in the early 1800s had more than a few years of formal education, and the design of complex structures was still the business of men who had learned on the job, mostly through knowledge passed along to them by their fathers and (to a sometimes dangerous degree) through their own trials and errors. At the turn of the nineteenth century the design of bridges in America was still a fairly primitive process, and the work was for the most part limited to relatively short spans. But a small crop of carpenter-builders had begun to emerge, led by three men who had learned their craft as apprentices: Timothy Palmer, Louis Wernwag, and Theodore Burr.

Palmer, born in 1751 in Rowley, Massachusetts, had apprenticed in his hometown and had spent his early years working on the design and construction of buildings (including at least one important church in Newburyport, Massachusetts) before turning to bridges. By 1800, he was building bridges that blended arches with trusses in new hybrid forms, first crossing the Merrimack River in Massachusetts and later crossing the Schuylkill in Pennsylvania. The Schuylkill crossing, called the Permanent Bridge, included a daring single span of 195 feet and two others of about 150 feet each, and would justify its odd name by remaining in service for ninety years before being torn down.

Louis Wernwag was born in Germany in 1769 and emigrated to Philadelphia in 1786. Like Palmer, he had learned his trade as an apprentice and had spent his early years working on designs for buildings, concentrating on mills and factories. During the early part of the nineteenth century Wernwag turned to bridges, and in 1812 he completed construction of a famous bridge near Philadelphia that carried vehicles and pedestrians across the Schuylkill River on trussed arches that were a spectacular 340 feet long.

At the Waterford-Lansingburgh crossing on the upper Hudson, it was Theodore Burr, the third of the three bridge pioneers, who would be engaged to design and build the bridge. Born in Torringford, Connecticut (later part of Torrington), in 1771, he had received a slightly better than average education before learning the carpenter's trade from his father, a well-established builder of mills and the operator of a grist mill of his own. Young Burr, who is said to have been a cousin of the infamous Aaron Burr, later married a granddaughter of the English navigator Captain James Cook, and in 1793 he had moved on to establish himself as a millwright and as a mill operator in the newly settled hamlet of Oxford, New York, along the Chenango River. In Oxford, he began his career in bridge construction by first designing and building one for himself, along with a small dam, to facilitate the operation of his mill. When the Union Bridge Company engaged him for its job at Waterford, he had completed or was in the process of designing several other bridges, including a substantial drawbridge across Catskill Creek at Catskill, New York, and an especially long arch bridge across the Mohawk River at Canajoharie.

It's not clear whether the proprietors of the Union Bridge Company knew before they engaged Burr that his Canajoharie bridge had begun to show serious signs of structural trouble soon after its completion. (It would collapse in 1807.) What is clear is that they must have had a strong sense that Burr was an extraordinarily capable fellow and destined to become one of the country's most distinguished and prolific bridge builders. Nor did the 1807 disaster in Canajoharie shake the confidence of the town fathers there; they engaged Burr to design and build the replacement bridge as soon as the debris from the failed one had been removed. Burr, who went on to design more than forty more major bridges during the next sixteen years or so (with many more successes than failures), was only thirty-two years old when he accepted the commission to design the Waterford-Lansingburgh Bridge in 1803.[8]

No stranger to the harsh winters of upstate New York, Burr began preparations immediately for starting foundation work in the spring of 1804, knowing well that the underpinnings of the bridge would be every bit as difficult and time-consuming as its superstructure. Excavation, dewatering, and material handling without power-driven equipment would make almost every process slow and difficult, exacerbating pressures already imposed by the short length of the working season.

The bridge company moved quickly to advertise for masons to build the bridge's stone piers and abutments, and by April a substantial workforce had been recruited and work was under way. The piers were to be relatively shallow, but their loads were great enough to require bases as long as 100 feet and as wide as 25 feet.[9] Isolating such excavations with crude timber cofferdams and then dewatering them with primitive, manually powered equipment would be demanding, tedious work, but by summer the pier and abutment work was well advanced and fabrication of the timber arches and trusses, which had been started right away along the riverbanks, was catching up fast. All that good progress attested to the industry and ingenuity of workers who had never known the luxury of power tools but were well acquainted with and skilled in the use of the block and fall, the hand pump, the screw pump, and the adze.[10]

Burr resolved the crossing into four river spans of slightly different lengths, ranging from 154 feet to 180 feet and producing a bridge just under 800 feet long. His design was essentially a traditional arch structure in which deep and heavy arches (doubled 8 inch × 16 inch timbers) were supported by massive stone abutments and piers, but it would feature Burr's newly developed and somewhat radical technique for integrating structural trusses to work in combination with the arches. His design provided for parallel chord timber trusses securely connected to the arches by oak pins driven through the approximately 10 inch × 10 inch verticals and diagonals of the trusses, effectively enabling the trusses to stiffen the arches and to resist their deformation as loads began to move across the bridge. There were three parallel rows of these big timber arches, dividing the almost level deck of the bridge into two approximately 15-foot-wide roadways and finishing flush with the tops of the trusses, a little less than 15 feet above the roadway.[11]

The work began to wind down toward the end of November 1804, just when winter was beginning to challenge the right of the builders to be working on the river. In only a little more than eight months, Burr and his crew had managed to complete both the foundations and the superstructure of an uncovered bridge that was almost as long as three modern football fields, a feat that would be regarded as significant even in later times. On the third of December the newly elected governor of the state, Morgan Lewis, led a celebration of the bridge's opening, and the directors of the Union Bridge Company enthusiastically toasted Burr and his principal lieutenants: Samuel Shelly, the master carpenter and general superintendent, and James McElroy, the head mason. A seventeen-gun salute was fired to honor the original thirteen colonies, as well as Kentucky, Ohio, Tennessee, and Vermont, the four states that had entered the Union since the end of the Revolution.[12]

The the Waterford-Lansingburgh Bridge would later seem a modest one, when compared with the railroad-carrying structures that would follow it,

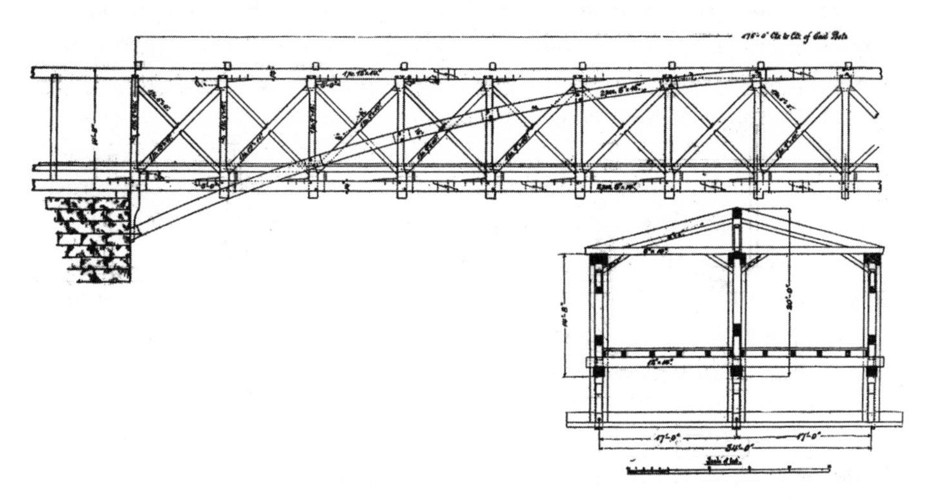

FIGURE 2 The superstructure of the Waterford-Lansingburgh Bridge. A drawing prepared in 1889 to show the details of the arch-truss construction used by Theodore Burr on this bridge eighty-four years earlier. It was prepared by the eminent engineer Theodore Cooper for a paper he read before a meeting of the American Society of Civil Engineers and has been made available here by the courtesy of the Linda Hall Library of Science, Engineering & Technology, Kansas City, Missouri.

but some writers (noting that its new use of the truss-arch combination was a historic advance in bridge technology) have described it as the greatest wooden span of its time. It endured moderately well, but, like all unprotected wood structures exposed to the elements, it suffered badly when variations in temperature and moisture affected its dimensions, opening its joints and in some cases impairing the health and durability of its members. Although timber bridges were often covered with wood sheathing in those days, the Waterford-Lansingburgh Bridge was not covered at first, probably for reasons of economy, and in 1812 the directors decided to revisit that decision. Between 1812 and 1814 the Union Bridge Company sold about $20,000 worth of additional stock and used the proceeds to remove and replace defective material, to cover the sides, and to roof the bridge with wood, giving the whole structure a new lease on life. With the help of regular maintenance and occasional metal reinforcement, the bridge remained in service for almost another century, but in 1909 a terrible fire, stoked by a gas line that had by then been attached to the bridge, destroyed the whole structure in only a few hours. It was replaced by a steel truss bridge that was erected on the original (repaired and reinforced) stone piers.[13]

Burr didn't survive as well or nearly as long as his bridge. He went on to build a good many more bridges, including a few that became especially well known for their extraordinarily long clear spans (one exceeding 360 feet) and some for their exceptionally good appearance. But several of his bridges were destroyed by ice or floods within only a few years of their completion, tarnishing his good reputation a little. Like many other men whose skills were essentially technical, Burr apparently wasn't good at managing a business, and about a dozen years after he had finished work on the Waterford bridge, at a time when he had several big jobs under way along the Susquehanna River, that weakness began to catch up with him. He was unable to pay his subcontractors, although funds earmarked for them had apparently been advanced to him, and his finances were described as being in disarray. A few years later he simply disappeared. Although it was said that he might have been working as a construction superintendent on a job in Pennsylvania, there doesn't appear to be any reliable verification of that report. Just what happened to him remains a mystery. The elegant home he had built before 1800 in Oxford, New York, was being used in 2008 as the town's library. His heirs report that he died in 1822, but they have been unable to say what caused his death or to identify the place of his burial.[14]

CHAPTER 2

Steam and a Bridge at Troy

THE GENUINE IMPORTANCE OF THE BRIDGE between Waterford and Lansingburgh notwithstanding, it's difficult to argue that it was a watershed event in the history of the Hudson Valley. Burr's early and effective use of the combined arch and truss (which led the editors of *Engineering-News* to describe the bridge in their editon of 1 June 1889 as "the greatest existing wood span in the world") gave it a unique place in the history of civil engineering, and its role as the first bridge to cross the navigable Hudson is clearly noteworthy. But in the context of the complex, developing history of the Hudson Valley, it was a structure that did not respond to or bring about any compelling or historically important change in the region. It could probably have been built a little earlier or later, or even in a different location, without significantly affecting the history of the region.

The railroad bridge at Troy is another story. Thirty-one years after the Waterford-Lansingburgh Bridge was finished, a bridge between the city of Troy and the village of Green Island, just a few miles downstream from Waterford, marked a change in the course of history in the Hudson Valley. The Age of Steam had arrived, bringing with it power for industry in the valley, speed and reliability for vessels on the river, and railroad connections to the rest of the country. The bridge at Troy was the first to carry a steam railroad across the Hudson, and as such it was an agent of historical change.

Not that steam power was a new idea in 1834. Scientists and other inventive people had been thinking about it for hundreds of years by then and had made some progress toward controlling and harnessing it. As early as 1698, an Englishman named Thomas Savery, building on the still fairly primitive work of several even earlier tinkerers, had been able to remove water from the depths of flooded English mines with a device that relied on the vacuum created when steam condenses. A couple of decades later, another Englishman, Thomas Newcomen, refined and further developed Savery's work to produce a single-acting, steam-driven piston engine that could pump out mines.

All that groundbreaking work and still other ideas were brought together and expanded during the late 1760s, when a singularly brilliant Scot named James Watt, working on repairs to a Newcomen engine at the University of Glasgow, developed and patented a significantly more efficient and more effective steam engine of his own. When he teamed up in 1775 with Matthew Boulton, an affluent English industrialist who had earlier worked with Benjamin Franklin on a steam engine, the seed for one of the Industrial Revolution's most important inventions was sown. By the end of the eighteenth century, when the patent for the Watt-Boulton engines expired, more than three hundred of them were already working in the mines and mills of Great Britain and Ireland.[1]

America, meanwhile, where the potential for applying steam power was as promising as anywhere in the world, hadn't yet entered the industrial age. Its foundries and small factories were becoming more productive, but there weren't many of them, and they were for the most part still focused on turning out the simple products that a largely agrarian population wanted. Steam power certainly was not central to their operations.

But a few forward-looking signs were beginning to appear in the Hudson Valley. Limitations on the ability to transport goods and people into and out of the area's counties were increasingly seen as major barriers to growth. Serious people were starting to look for a way to improve the situation, and the more they heard about steam-powered railroads, the more interested they became. The trouble was that the center of the relevant technology was in Great Britain, and the cost of bringing such a complex new technology and its equipment across the Atlantic seemed beyond anything the Americans could afford. By the end of the eighteenth century, somewhat discouraged, these entrepreneurs had pretty well accepted the idea that if the new Republic was going to move effectively into its first full century, they were probably going to have to figure out a way of developing such a technology for themselves.

Steam power was certainly not unknown in America at the end of the eighteenth century. Although Americans had not until then paid much attention to its potential application to railroads, by about 1800 a few steam engines were being used in the country's mills and mines. In those years the British government still barred the export of engines, so anything used in America had to have been cobbled together by enterprising home-grown mechanics or by the occasional skilled technician who had left England with the details of the latest in steam engine technology sewn into the lining of his jacket or preserved in his memory.

By this time, more than a little attention was being focused on the use of steam power for boats, a direction that would profoundly affect the development of the Hudson Valley. A Maryland-born wheelwright named Charles Rumsey had already produced a steam engine that had propelled a small boat down the Potomac River by emitting a strong jet from its stern. But he couldn't get enough financial support to continue his work in America, so he moved to England, where he died penniless a few years later.[2] John Fitch, a crusty, resourceful, and clearly gifted mechanic and inventor, didn't fare much better. Around 1790, using a popular British publication as a guide, he built his own steam engine and installed it on a boat of his own, where it drove an endless chain attached to a set of paddlewheels. At six knots, his steamboat was said to have passed every other craft on the Delaware. But like Rumsey, he failed to get funding and died in Kentucky in 1798 of an overdose of opium.[3]

Other men threaded their way into and out of the business of developing steam power for boats during the last years of the eighteenth century and the early years of the nineteenth. The most successful of them wasn't an inventor at all. He was Chancellor Livingston, the scion of a wealthy New York family and a powerful figure in the state's political life. He teamed up with John Stevens, a capable inventor who was having some success with steam engines and was married to Livingston's sister. Livingston was able to persuade the State of New York, which was eager to encourage the development of steam power, to promise the Livingston-Stevens partnership an absolute monopoly on all steam navigation on the Hudson if it brought viable steam transportation to the river. The partners split after a few unsuccessful years, but Livingston retained control of the monopoly and did not give up.

What Livingston needed was a new partner, and he found one in 1802 in the person of Robert Fulton, a young, temperamental artist and inventor who was looking for a new direction in his own career. Livingston promised him financial and political support and half the proceeds of the monopoly if he could produce a workable engine. Five years later, near the mouth of the Hudson, Fulton launched a 114-foot long, steam-powered boat called *North River* (later called *Cleremont*). It was a historic event. After twenty-seven hours of travel under steam power alone, much of it against both wind and tide, Fulton's vessel completed the 135-mile voyage to Albany.[4]

The treasured monopoly was awarded as promised, but the revolution in river travel that Livingston and Fulton had counted on didn't materialize. Few skippers on the river were willing to pay the fees demanded by the Livingston-Fulton partnership and simply stuck with their sailboats. Very little attention was paid to steam power on the Hudson for more than another ten years, until a scrappy sea captain named Thomas Gibbons decided to challenge the monopoly in the early 1820s, claiming it was unconstitutional. Gibbons had been operating steam-powered vessels under an arrangement with a shipowner named Aaron Ogden, who had knuckled under to the monopoly and was paying their fees. Ogden was passing the burden of the fees along to Gibbons, who soon tired of them and refused to continue the arrangement. He argued that control of navigation on inland waterways like the Hudson was the exclusive business of the federal government and that the state had exceeded its authority in granting a monopoly to Livingston and Fulton. The Supreme Court of the United States agreed with him, and the monopoly collapsed. That decision, made in 1824, marked the real beginning of the Industrial Revolution in the Hudson Valley. Before long, the river would come alive with steam-powered vessels, and many towns along the Hudson's banks would begin a long history of growth and prosperity.[5]

The collapse of the monopoly allowed the steam engine to revolutionize river transportation, but that was only the beginning. In 1825 further events pointed to steam-related changes in land transportation as well, and these would be every bit as radical as the changes on the river. During that eventful year the Erie Canal was completed, making it possible for vessels to travel by water all the way from Buffalo to Albany, unimpeded by rivers, swamps, or hills, or even (thanks to an effective system of locks) by the 675-foot difference in elevation between those two cities. The canal's cargoes of eastbound products from the farms of what was in the nineteenth century considered the western United States would soon lead to prosperity and population growth in the region that produced them and to the statehoods of Ohio, Illinois, and Indiana. Cargoes of westbound manufactured products would further enhance the prosperity of the Hudson Valley, and most especially of New York City. All of what was happening would intensify traffic along the Hudson itself, encouraging the use of big, steam-powered vessels and propelling the whole valley into a new phase of growth. Steam-driven railroads were not far off, and most of them were going to require bridges.

Not everything on the great new Erie Canal was perfect. In addition to freight, the canal carried passengers, and for them the easternmost leg of the canal journey, the leg that connected the inland city of Schenectady with the state capital at Albany, was a tedious one. Although the direct distance between those two cities was only about seventeen miles, geologic conditions had required the canal to follow a circuitous path almost twice as long. For the eastbound canal boat passenger who wanted to avoid that unpleasantly tedious last leg by disembarking at Schenectady and traveling overland directly to Albany (or for the westbound passenger considering the overland option from Albany to Schenectady) there was a potential saving of ten miles or so. But the road was rough, and the trip by stagecoach could take at least a couple of hours. There was talk about connecting the two cities with a railroad along the short, direct route between them, but steam railroads were still a new idea in the United States in 1825, and it would be another five years before there would be an operating steam railroad anywhere in the country that could provide a decent model to guide planners in New York.

One man in Schenectady who thought it made sense to build a steam-powered railroad between his town and Albany was George Featherstonehaugh, an English-born American who lived near Schenectady (and whose difficult name became mercifully shortened in conversation to Fen-shaw). He was aware that the development of the steam railroad was much farther along in England and that only a few months before the opening of the Erie Canal it had taken a dramatic leap forward. A steam-powered locomotive of the Stockton and Darlington Railroad, with its designer George Stephenson at its controls, had hauled thirty-six heavily loaded wagons a distance of nine miles in about two hours.

Featherstonehaugh recruited a few partners—among them, a rich and politically powerful scion of the Van Rensselaer family—and they were able to secure a state charter in 1826 for incorporating the Mohawk and Hudson Railroad Company to run between Schenectady and Albany. (Thirty years later it would be the first link in what would become the vast New York Central system.) Financing and construction took longer than expected, but by the middle of 1831 steam-powered locomotives of the Mohawk and Hudson were pulling trains between Schenectady and Albany.[6]

Not every turn in history has profound roots, and certainly the one that led to the construction of an additional twenty-three miles of railroad farther north beyond Schenectady, to Saratoga Springs, did not. What precipitated that extension was little more than the taste that Americans had acquired around the beginning of the nineteenth century for the mineral waters that bubbled up naturally at and near Saratoga Springs and a few miles south of Saratoga Springs at Ballston (later called Ballston Spa). By about 1831 the demand for what were touted as the health-giving properties of such waters had led to the construction of numerous hotels and boardinghouses, a burgeoning commercial infrastructure, and a seasonal surge of travelers that was seen by investors as more than enough to justify the cost of extending the new railroad from Schenectady to Saratoga Springs. The Saratoga and Schenectady Railroad, incorporated as a steam railroad in 1831 and completed by 1832, lengthened a continuous steam railroad system that extended all the way from Albany to Saratoga Springs, via Schenectady and Ballston Spa. The steam railroad had come to New York State.

Just about everyone in the region was happy about the new railroad except the merchants of Troy, the Hudson River port located a few miles north of Albany on the opposite side the river. Troy merchants had reasonably expected that the Erie Canal would bring them a big share of the new prosperity, much of it across the bridge that since 1804 had linked the towns of Waterford and Lansingburgh, Troy's northerly neighbors. But now this new direct rail shortcut that the Saratoga and Schenectady had built between Albany and Saratoga Springs threatened to draw canal traffic away from Waterford and to deliver it directly to Albany, without giving passengers a chance to cross the river to Troy. The Troy merchants favored building a railroad of their own to connect with the Schenectady and Saratoga at Ballston Spa, making it easy for passengers on that new system to choose to travel to or from Troy. Such a link, they figured, would siphon off a fair amount of the Albany-bound traffic.

A couple of enterprising Troy businessmen recognized this opportunity to bring business to their town, a place they knew and understood especially well. One was Stephen Warren, a prosperous member of a family that had been in the colonies since 1655 and that had relocated to Troy in 1798. Warren was the owner of a successful stove manufacturing business and a principal in several other local businesses. The other businessman was Richard P. Hart, who would become mayor of Troy within a few years. Hart had accumulated a modest fortune as the proprietor of stage lines he been operating for years between Troy and Lake Champlain until the new Champlain Canal, completed at about the same time as the Erie Canal, displaced the stages. These enterprising fellows had the inclination and wherewithal to invest in a new railroad venture, and in 1834 they brought in a group of like-minded locals to establish the Rensselaer and Saratoga Railroad, named for the two counties it would connect. The consortium quickly set about laying the groundwork for a railroad that would start in Troy, travel north along the east bank of the river to Lansingburgh, then cross the Hudson on the bridge between Lansingburgh and Waterford, and continue west and north to connect with the Saratoga and Schenectady at Ballston. The fact that the 1804 Waterford-Lansingburgh Bridge had never been designed to carry railroad loads apparently did not discourage them, and it can probably be assumed that they anticipated some need to reinforce the bridge.

What this seasoned group of Troy businessmen apparently had not counted on was the price their upstream neighbors would demand for the right to use the Waterford bridge. There's some evidence that the bridge's proprietors thought the language of the new railroad's charter would force the developers to use the Waterford bridge, but the Troy group thought otherwise. They moved quickly to secure a legal opinion that supported their right to avoid those high fees at Waterford by building their own bridge at Troy.

Troy's rivals downstream in Albany, who already resented the 1804 bridge at Waterford as an affront to their more commercially advanced city, had been engaged in an ongoing but unsuccessful campaign since 1814 for the right to build their own bridge, so they didn't take the plans of the Troy developers lightly. They protested wherever and whenever they could, but to no avail. By 1832 Hart and his partners in Troy had secured authorization from the state legislature to build a new bridge between Troy and Green Island, on the Hudson's west bank. It would be the first railroad bridge to cross the Hudson.

Developing a workable plan for a new bridge at Troy did not come without problems. Originally, the sponsors had planned to run their railroad up the east bank of the river before turning west at Lansingburgh to cross on the Waterford bridge. To cross at Troy would mean running trains up the west bank of the river instead. That would be an especially expensive change, because it meant crossing four "sprouts" of the Mohawk River along the way via four short but expensive new bridges (three of which would burn down, in separate fires, in the next few years). In addition, the river at the Troy crossing was almost twice as wide as it was at Waterford, making the bridge more than twice as expensive to build. Furthermore, the need to allow tall vessels to pass under or through it—a problem not faced upstream, where the bridge was north of the navigable limit of the river—would have to be addressed for the first time. And just to make things more difficult, Theodore Burr, the locally experienced designer of the 1804 bridge, had died, leaving the Rensselaer and Saratoga group to start from scratch in their search for a bridge builder.[7]

They began by engaging Isaac Damon of Northampton, Massachusetts. Unlike Burr, whose experience had been mainly in carpentry, Damon had gone on from carpentry to an education in the design and construction of buildings as an apprentice to the distinguished Boston architect Asher Benjamin. Damon later started his own, independent career almost accidentally while still working under Benjamin. During the construction of Benjamin's Northampton Meetinghouse, the contractor on the job defaulted, and Damon took on the task of supervising the rest of the work. His good performance attracted the attention of public officials in Northampton, and after the meetinghouse was finished, in 1812, direct commissions to design and build a couple of nearby public buildings followed, launching Damon's career. He went on to design and build a substantial number of extremely well-regarded churches in Springfield, Blandford, and Greenfield, and more than a few elegant residences, some of them still extant in the twenty-first century and well preserved as landmarks. Along the way, Damon added to his practice the design and construction of bridges, and by the time the Rensselaer and Saratoga Railroad job came along in 1833, bridges had begun to dominate his work. By the end of his life, in 1862, Damon had designed fifty-two structures, of which twenty-five were bridges.[8]

A practical man, Damon did not believe in reinventing the wheel. For the principal structural element of his new bridge at Troy, he elected to use a truss that had been designed and patented by an exceptionally talented contemporary named Ithiel Town. The Town truss was no ordinary one, and its designer was no ordinary man. Like Damon, Town had been born in 1784, and, like Damon, he had learned the craft of architecture under Asher Benjamin of Boston, not far from his own northeastern Connecticut roots. Once out of his apprenticeship, Town had quickly distinguished himself with designs for Center Church and Trinity Church on the New Haven Green, and soon after those buildings became known and admired, his services began to be sought throughout much of New England and New York. In the years that followed, he would design (alone or in collaboration with other architects) state capitols in three cities and the especially distinguished U.S. Custom House in New York, as well as a wide variety of residences, churches, and other public buildings. His reputation as one of the most important architects of the period would be enhanced years later when Samuel F. B. Morse selected Ithiel Town to be one of only two architect members in the newly formed National Academy of Design.[9]

In addition to sharing with Isaac Damon his birth year and his place of apprenticeship, Town shared his interest in the design and construction of bridges, and early along that interest began to show up in his practice. He felt as strongly as Burr that more reliance should be placed on the structural truss than on the structural arch in the design of new bridges, but he went one step further. Town joined a small but growing number of bridge builders who were abandoning the arch entirely and relying on the capacity of properly designed trusses to carry loads.

Among the arguments that supported such a radical approach, two were especially compelling. One recognized that, unlike the arch, which required massive piers capable of resisting not only the vertical loads of the bridge but also the horizontal thrust of the arches, the trusses would impose only vertical loads; therefore, the piers could be relatively narrow. Second, the wooden arches popular in America, unlike their massive masonry forebears in Europe, were usually thin and light, lacking the stiffness that could be obtained with trusses.

His advocacy of trusses did not blind Town to their weaknesses, and he addressed those vigorously in his designs. He realized that the tedious work of finding and fashioning the heavy timbers normally required for the diagonals and chords of conventional trusses (like Burr's) slowed the progress of construction and increased its cost. In addition, he realized that the intricate joinery needed for critical connections between and among the heavy members required the skills of the most experienced craftsmen, further increasing cost and construction time. To deal with these issues, Town designed and in 1820 obtained a patent for the Town Lattice Truss, in which an exceptionally large number of light, closely spaced diagonals were connected in a lattice configuration and rigidly secured to relatively light top and bottom chord members. The finished truss then consisted of multiple diagonals assembled in two adjacent planes: an outboard plane had parallel diagonals placed at an angle of about 45 degrees with the horizontal, all facing one way; and an inboard plane had similar diagonals, all parallel to each other and all making the opposite 45-degree angle with the horizontal. The finished truss appeared from the side as a continuous assembly of diamond-shaped openings, bordered by top and bottom chords, with the diagonals connected to one another and stiffened at every intersection by a couple of inch-and-a-half-diameter oak treenails, driven snugly into augered holes that obviated costly mortise and tenon work.

What especially endeared the Town Lattice Truss to the people charged with building new bridges was the ease with which it could be fabricated at a sawmill. Nearly every wooden diagonal was an identical 3 × 12, about 24 feet long, leaving only the roof and floor beams, the chords, and a few special pieces to require heavier timbers. What Town had designed was a serviceable structural truss that could be easily fabricated by workers of limited experience and was capable of supporting any loading that could be reasonably expected from early nineteenth-century railroad traffic. When the occasional longer span or heavier loading was anticipated, Town's trusses could be doubled up. They were quickly accepted wherever in the country new bridges were being built, and their popularity would continue through the middle of the century, when dramatic increases in spans and loads would begin to render them obsolete.[10]

Damon brought in a builder named Joseph Hayward to manage the construction work in the field, and by early summer in 1834 work on the stone piers and abutments was underway. The eight stone piers and two stone abutments would divide the approximately 1,512-foot-long bridge into nine sections. Documents detailing the construction of the piers are no longer available, so it is not known whether or not piling was required to ensure the support of bedrock below the river.

Although the builders of the Waterford-Lansingburgh Bridge had been able to ignore the issue of headroom under their bridge, the builders of the Green Island Bridge had to provide a way for ships to pass under or through it. They chose the latter option: one section of the bridge, closer to Troy than to Green Island, was built as an approximately 60-foot-long, swing-style draw section. Each of the other eight bridge sections was about 180 feet long, a span for which the Town Truss was well suited. The stone work for the piers and abutments was finished by the time the 1834–1835 winter weather set in, and by spring everything was in readiness for the timber superstructure.

Construction of the bridge that would span the approximately quarter-mile-long gap between the village of Green Island and the city of Troy was begun in the early months of 1835. Before the end of the year two enclosed, 17-foot-wide travel lanes would be completed, one for horse-drawn carriages and wagons and the other for a combination of railroad tracks and a pedestrian walkway, separated from one another by a low railing. The finished structure would be typical of the covered bridges of the period, with a pitched, shingled roof about 17 feet above the wood plank flooring at its eave line and with its exterior trusses sheathed with wood boards. Some (but not much) light would be admitted to the two darkened aisles through thirty-two hinged skylights. Where the bridge crossed Center Island, an approximately six-foot-wide opening was left in one of its outside walls to allow workers employed by the Starbuck Machine Works, whose shop was located there, to enter and leave.[11]

Oddly enough, there's not much evidence in contemporaneous periodicals that the people of Troy paid much (if any) attention to the extraordinary achievement represented by the completion of this first railroad bridge to cross the Hudson. When it was finished, there was no speech by Governor William L. Marcy, although he was a Troy native, and apparently not even so much as a marching band. But very early in October 1835 a small advertisement for the Rensselaer and Saratoga Railroad appeared in the *Troy Daily Whig,* announcing that service between Troy and Ballston had been initiated and that passengers would be accepted at the new Rensselaer and Saratoga station in Troy every morning at 10 a.m. The advertisement looked at first like something intended to announce the opening of the bridge, but it wasn't. It included a revealing disclaimer that such passengers would be first taken from Troy to Waterford by a stage that would cross the river by way of the old bridge between Lansingburgh and Waterford, and that only when they got to Waterford would they board a train for Ballston. Clearly, the new bridge between Troy and Green Island was not yet really finished. It would be a few more weeks before the Rensselaer and Saratoga announcement would be modified to omit the disclaimer, indicating that the new bridge between Troy and Green Island had indeed been completed and was fully operational.

Even then, the cars of arriving trains would be separated from their locomotives at Green island and drawn across the new bridge by teams of horses, while the locomotives remained idle at the bridge's west end. Fearful that an errant spark from one of the locomotives might find its way into the timbers of the new bridge, the town fathers in Troy had denied permission for any locomotive to cross the bridge.

What must have seemed to many like an unnecessarily conservative restriction would remain in effect for almost twenty years, until the bridge was widened by the addition of a parallel span. In 1854, in what was widely viewed as a victory for progress, steam locomotives were finally permitted to use the widened bridge, although even then they were allowed to travel only on the added structure, leaving the original bridge for the use of carriages and pedestrians.

Eight years later, the city fathers of Troy had reason to regret their flirtation with progress. The once-feared spark from an engine did indeed ignite the timbers in the bridge, creating the most catastrophic fire in Troy's history. It destroyed the easternmost sections of the bridge and—driven by strong winds—spread into the city of Troy itself, taking five lives and destroying 507 buildings over a seventy-five-acre area.[12]

The fire of 1862 was a wrenching disaster by any measure, but by the time it occurred, the bridge had become an integral part of the transportation network of the region and there was no hesitation about replacing it as quickly as possible. Twenty years earlier, another railroad had been added to the systems that already relied on the bridge at Troy, increasing public dependence on the bridge and enhancing Troy's developing commercial relationship with the burgeoning western parts of the state. The work of replacing the destroyed section, essentially by replicating the original design, was begun almost immediately after the fire's debris had been cleared. Although there was some support for the increasingly popular use of wrought iron for the new trusses, a decision (probably influenced by the need for speed) was made to build the replacement in wood. Fourteen years later, the original wood sections that had survived the 1862 fire without damage would be replaced by iron trusses, and in 1884 the timber sections that had been erected immediately after the fire would themselves be replaced by iron trusses. What had come to be known as the Green Island Bridge (it had originally been called simply the Rensselaer and Saratoga Bridge) became a modern wrought-iron structure.

The railroad systems that relied on the Green Island Bridge would later be absorbed into the Delaware and Hudson Railroad system, and not long after the turn of the twentieth century, as rail traffic declined and automobile traffic increased, the future of the bridge as a rail crossing became increasingly tenuous. Rail service at Troy eventually ended, and in 1963 the bridge was given over to the exclusive use of highway vehicles. In the spring of 1977 abnormally high river flows created by torrential rains and an extraordinary volume of melting snows combined to undermine one of the bridge's piers, shifting its position, collapsing a section of the roadbed, and destroying the steel bridge. A little more than four years later, a sleekly modern lift bridge that seemed to belong to an era that had not yet arrived was built in its place, and it would still be operating in the early twenty-first century.[13]

CHAPTER 3

Three Railroad Bridges at Albany

FOR A WHILE, THE HISTORIC BRIDGE that was completed in Troy in 1835 provided the only way a railroad train could cross the Hudson twelve months a year, regardless of weather or ice in the river, and the town's emergence as the prospering industrial and commercial center of the region owed much to it. By the middle of the century, Henry Burden, whose iron shop in Troy was producing (among many things) almost a million horseshoes a year, was running his plant with a huge water wheel that was generating 300 horsepower, and a nearby Methodist preacher had begun the local manufacture of detachable shirt collars, establishing an industry that would later become the largest employer in Troy. The population of the town had reached 30,000, and the small plants that abounded there included sawmills, iron forges, and breweries; factories for making stoves, carriages, and rope; and a plant that manufactured railroad cars. The prestigious Emma Willard School and the Rensselaer Institute (predecessor of RPI) were well established.[1]

Six miles downriver, the town fathers in Albany had been watching that growing prosperity in Troy with an interest tinged with resentment. They had felt for years that Albany, not Troy, should have been the location of the first railroad bridge across the Hudson, and they had been trying for a long time to get the state to authorize them to build their own bridge. But a determined coalition of Troy businessmen and influential ship owners had been able to frustrate their every effort, managing to defeat a formal petition that Albany submitted to the legislature as early as 1814 and continuing to deflect subsequent petitions every few years thereafter.[2] By mid-century the Troy faction that opposed an Albany bridge had even won some support within the city of Albany itself, recruiting the carters who were hauling passengers and freight to and across Troy's bridge whenever the Albany ferries were delayed or shut down. The number of people with axes to grind was formidable, and increasing.

But there was something else developing during those years that would eventually prove to be more persuasive in the halls of the New York State Legislature than any argument the bridge opposition could muster: it was the steam railroad itself. It had fully arrived and was taking the country by storm. Once that little railroad of George Featherstonehaugh's had begun to operate its locomotive between Albany and Schenectady in 1831, the idea of extending a steam-powered rail system all the way across the state, from Buffalo to Albany and continuing east from Albany toward Boston or south to New York, began to be taken seriously.

A section between Schenectady and Utica had been quick to follow the Albany to Schenectady line, and over the next dozen years or so an assortment of other short railroads sprang up to connect Utica with Syracuse, Syracuse with Auburn, Auburn with Rochester, and Rochester with Buffalo. It was a bumpy ride, fraught with hazards and replete with gaps and difficult transitions, but it was the crude beginning of the continuous steam railroad between Albany and Buffalo that was the precursor of the great New York Central Railroad.

What that loosely organized chain of railroads and their owners needed most was someone who would bring them all together into a single, smoothly operating system, and in 1853 they found just the man. He was one of their own: Erastus Corning of Schenectady, an autocratic, self-made industrialist and formidable political leader who had by then been successfully running the Utica and Schenectady (U&S) Railroad for almost twenty years. Even though, during his long tenure at the U&S, he required that the shareholder-owned railroad purchase all rails, tools, and related materials from the supply company he owned, Corning continued to be well regarded by his board and by the growing industry of which the U&S was a part. He had no trouble establishing a new board of capable and sometimes controversially loyal executives for the extended railroad system he was building. In short order he consolidated all the railroads along the route between Albany and Buffalo, devising occasionally complex and awkward linkages, easing their transitions, and ultimately establishing a rail system second to none. With a capitalization of almost $25 million, the Central would become (and remain, for a while) the largest corporation in the United States, and it wouldn't be long before the system would render obsolete the still relatively young and widely admired Erie Canal, changing the history and culture of the region forever.[3]

All that activity was occurring west of the Hudson. At about the same time, east of the river, a plucky group of entrepreneurs created the Western Railroad (later called the Boston and Albany), which was driving tunnels, building bridges, and laying track between Greenbush, the town just across the Hudson from Albany that would later be called Rensselaer, and Boston. With the Central operating between Buffalo and Albany, completion of the Western between Greenbush and Boston would provide continuous rail capability all the way from Buffalo to Boston. All it really needed was a bridge across the Hudson at Albany to connect the western section to the eastern section.

The commercial and industrial potential of such a system certainly wasn't lost on the rest of the Hudson Valley, least of all on the flourishing commercial center of New York City. What had started down there a few years earlier as a horse-drawn railroad to connect the city's center with the rural village of Harlem, about seven miles north, had by 1852 become the steam-driven Harlem Railroad. Having acquired the charter rights of the New York and Albany (a railroad that never got beyond the charter stage), the Harlem soon extended its tracks northward beyond its original destination into Westchester County, past White Plains, and on through Putnam and Dutchess counties into Columbia County, a total of 137 miles from its starting place. Mindful of the pitfalls of trying to build along the Hudson's edge (and eager to avoid being seen as a direct competitor of the powerful steamboat crowd), the Harlem's management maintained an inland course at a respectful distance from the river. At the Columbia County hamlet of Chatham, only about 23 miles southeast of Albany, it terminated its line and connected with the system that the Western was bringing across from Boston toward Greenbush, giving that increasingly important railroad a valuable connection with New York City and enhancing the Harlem's own access to the promising rail network that was taking shape upstate.[4]

And that wasn't all there was to this emerging network of railroads that was converging on the Capital District of New York. By the time the Harlem had made that important connection with the Western at Chatham, still another new railroad had been established by a group of substantial and ambitious Poughkeepsie businessmen who had surprised almost everyone with their skill and speed in bringing tracks of their own up the daunting east bank of the Hudson, parallel to the Harlem's line but virtually on the river's edge. Starting construction as late as 1847, the new Hudson River Railroad brashly ignored the bluster of the steamship owners and pushed on for Albany, driving tunnels through granite, bridging the numerous streams that punctuate the river's eastern shore, and laying a double set of tracks all the way. Ignoring the threats of steamboat operators and the fury of outraged Hudson Valley gentry who feared for the survival of their grand riverfront estates, the Hudson River Railroad had managed by the early weeks of 1851 to bring its system all the way from New York to Greenbush, terminating there alongside the Western.[5]

With the Central having long since established its line from western New York to Albany, on the west bank of the Hudson, and with both the Western and the Hudson River railroads now terminating directly across the river, on its east bank, the case for building a great bridge at Albany quickly gathered momentum. Late in 1851, the three railroads that would rely on such a bridge joined forces to establish the Hudson River Bridge Company, with the Central holding half the shares and the other two railroads dividing the remaining shares equally.

The next step was a campaign to secure from a historically reluctant legislature its authorization to build a bridge across the Hudson. Erastus Corning delegated that critical responsibility to a vice president named Dean Richmond, a rough-edged railroad man who had started his working career at the age of seven and was a passionate advocate of the bridge. A minimally educated physical giant of a man, whose handwriting was said to be so bad that only his signature could be deciphered, Richmond was a fellow of superior intelligence and persuasive style, and his task proved to be easier than expected. The political forces that had blocked earlier efforts yielded to the bridge proponents, whose base had deepened and broadened. Richmond now had the support of political leaders from all those counties and towns that lined the Central's long route across the state, each of them eager to ensure that the promised prosperity would be shared by his constituency. He had little trouble convincing them that what was needed was a bridge at Albany, and in April 1856 the legislature endorsed his proposal to build one. The Hudson River Bridge Company was authorized to proceed with the business of "erecting and maintaining a bridge . . . for railroad travel and transportation, across the Hudson River, from some point on the westerly side or shore of said river, in the City of Albany, to some point on the opposite side, in the County of Rensselaer." The bill went on to specify some of the characteristics of the bridge (not many), including its height above the level of common tide water (at least 20 feet), and to specify that a draw section must be provided, "of sufficient width to admit the free passage of the largest vessels navigating the river."[6]

The celebrations of the victors didn't last long. Only a few months later, the opponents of the bridge were in court seeking an injunction to prevent the state's approval from being implemented, and some of their arguments had enough merit to trouble the bridge advocates. The opposition contended that the state had exceeded its authority when it approved a potential obstruction to free navigation along the waterway, in violation of the federal Constitution, and that the whole issue of bridges across navigable waterways was the exclusive business of the federal government. Its petition argued further that the volume of river traffic was so high as to require opening the draw every five minutes, that a draw of such great size could not be opened or closed quickly enough, and that it thus would have to remain permanently open, rendering the bridge a useless obstruction. Even if that problem were to be remedied, those opposed to the bridge argued, natural changes in the topography of the silty bottom of the river, over the years, would affect the shipping channel in such a way that the draw would become useless. Furthermore, the petitioners wrote, the impact of even minor delays (like those imposed by the mere presence of piers) would be exacerbated for vessels whose progress was significantly affected by tides; such vessels would in many cases have to await a new tide cycle before resuming their progress. And just in case those substantial objections didn't do the trick, it was argued that so little information had been given about the details of the bridge's design or about its exact location that it was not really possible to evaluate fully its potential impact on river navigation.[7]

Of course, the bridge advocates had their lawyers, too, including no less distinguished a figure than William Henry Seward, a former governor of New York, later a senator from New York, and destined to become Abraham Lincoln's secretary of state. In September 1858 Seward presented the supporters' thoroughly reasoned and effective defense. The draw, they argued, could be operated by steam (although, in fact, it would not be) and could be opened or closed in only two minutes (which it would be). They said they would maintain the channel regularly and properly to prevent its wandering from the location of the draw (they did) and that they would position steam tugs at the bridge to assist any vessel that got into trouble (they did that, too). Seward emphasized his view that the constitutional right and duty of the federal government to regulate commerce on inland waterways did not in any way prevent the states from building bridges across them, so long as the bridges were designed, constructed, and maintained in ways that would ensure free navigation. His defense failed to address the complaint that very little information had been given about the actual design of the bridge, so there was no way to be sure that his clients would do everything they promised, but apparently that didn't bother the judge. He sided with the bridge company, refusing to issue the injunction.[8]

The forces opposed to a bridge at Albany reacted to the decision by pushing on to a higher forum, the U.S. Circuit Court. By this time, questions of fact had to a great extent been swept aside, leaving mainly the difficult constitutional question of just where the authority of the state had to yield to the acknowledged responsibility of the federal government to maintain the navigability of the country's waterways. The opposition continued to argue that the line had been crossed when New York authorized construction of a bridge across the Hudson at Albany, but the Hudson River Bridge Company stuck to its position that the existence of the bridge would not in any material way obstruct free navigation. In fact, even the court had trouble deciding just where the line was. As was customary in those days, a "circuit-riding judge" of the U.S. Supreme Court was brought in to sit alongside the Circuit Court judge, and on 27 January 1862 the two judges reached opposing conclusions about the Albany bridge, one favoring the injunction and the other opposing it. Protocol then required passing the case on up to the U.S. Supreme Court itself. When the justices of the high court found themselves equally divided, the lower court judge's refusal to order the injunction became automatically affirmed. The Albany bridge had won by a whisker.[9]

There was still plenty to be done before the first signs of construction would appear along the Hudson at Albany, and it would be several months before the activities of the bridge company would acquire any real momentum, suggesting that the favorable outcome of the lawsuit might have surprised the investors a bit. But within a year boring rigs were probing the river's bottom, property was being acquired on both sides, and there were indications that design work was underway. As soon as the ice melted in 1864, half a century after the first petition had been presented to the legislature and eight years after the legislature had first authorized the bridge, construction began on a bridge to connect Lumber Street (later called Livingston Avenue) in Albany, with a landing on the opposite bank in what would by the end of the nineteenth century be called the town of Rensselaer.

Bridges were nothing new for railroads, of course, and design and construction appear to have been placed in the hands of persons who had already been in the employ of one of the three railroads that owned the Hudson River Bridge Company. George E. Gray, who was the Central's chief engineer, took charge until the piles and pile caps were completed; he was briefly succeeded by Colonel Julius W. Adams, who does not appear to have taken a significant role. Once the serious work of building the stone piers started, a man appropriately named Bridgeford was put in charge; but in the long tradition of projects for which authority rests with multiple owners, enough dissatisfaction with the slow progress of the work developed to end Bridgeford's tenure. He was replaced by Alfred F. Smith (not to be confused with either of the other prominent Alfred Smiths who would later figure in the history of the New York Central and of New York State), a capable veteran of bridge work who was at the time of his selection the general superintendent of the Hudson River Railroad and well suited to the task. Under his strong leadership and supervision, the whole project would be completed within only two years, an impressive achievement in New York's difficult climate and in a period when most of the time-saving equipment of later periods had not yet come into use.[10]

The foundations, never easy on a bridge that crosses a busy river, comprised twenty-one massive stone piers, each supported on timber piles driven well into the river bottom and capped with concrete. Although there's no evidence that steam hammers were used for driving piles at that still early date, elaborate pulley techniques (often learned by carpenters who gained experience on sailing vessels) are known to have been widely used for raising and dropping rams and weights for pile driving, and it's likely that such techniques were used at Albany. Timber caissons and still rudimentary pumping systems kept the holes dry long enough to allow for placing concrete for the pile caps and for laying up the first few courses of stone. Most of the stone was brought upriver by barge from quarries near Kingston, and the solid masonry piers for which they were used were for the most part (depending on location) about 7 feet wide, more than 30 feet long (in the direction parallel to the stream) and about 60 feet high. The great height of the piers ensured almost 30 feet of clearance between mean low water and the bottom of the bridge structure (more than the charter had required), and their length had been calculated to allow for a possible future widening of the bridge deck to accommodate a second set of tracks.

One of the few departures from the almost uniform size of the piers occurred at the center of the draw section, where a single massive pier was built to support a giant turntable that would open and close a 230-foot draw section. The pier itself was about 260 feet long, perpendicular to the long axis of the river, and about 32 feet wide. As to the movable draw section itself, one of its two approximately 115-foot-long trussed arms was designed to open by swinging upstream, while the other was designed to swing downstream, producing two generous shipping channels that would allow for the simultaneous passage of two extraordinarily wide vessels. The turntable mechanism for opening and closing the draw was ingenious and elaborate, and was the subject of admiration and wonder because it did indeed allow the bridge to be opened or closed (even before a steam engine was added) in less than two minutes.[11]

By the onset of winter, late in 1864, foundation work was pretty well finished and ready to receive the superstructure, and the town was more than ready to see it. But what greeted spectators in the spring of 1865 disappointed most of them. Although the age of iron bridges hadn't yet begun in earnest, many of the important new bridges in America were already being built of wrought iron and cast iron, and most people in Albany thought this one would be, too. The New York Central had already built a number of iron bridges along its routes, and its largest competitor, the Baltimore and Ohio, was by 1865 boasting of having no fewer than seventy iron bridges in place. Only a few years later, the Lehigh Valley Railroad, another rival, would build a single-span, 165-foot-long iron truss railroad bridge across the Delaware River. The Hudson River Bridge Company had avoided providing construction details in advance, and under the circumstances most people had expected that the new bridge would reflect "state of the art" technology.

What they got was a bridge made of wood. The main superstructure employed a system of Howe trusses made of heavy timber, with top and bottom chords and diagonals rarely smaller than 12 inches by 12 inches and with iron verticals and a deck framed in heavy timber. The bridge's longest spans, where the trusses were about 24 feet high, measured about 172 feet long, and the sections that were closer to the river's edges, where the piers were outside the route of the ship channel and could be more closely spaced, were carried by shallower wood trusses.[12]

The use of wood instead of iron as the principal building material wasn't the only surprise. The oversized piers had led a good many people to expect a two-track system, but they were to be disappointed. The deck of the new bridge was too narrow for two sets of tracks, and what was taking shape in 1865 was a single-track, wooden bridge on which trains would be able to proceed in only one direction at a time, leaving trains bound in the other direction to wait. The whole grandly imagined thing was a disappointment.

The bridge company's managers made it as clear as they could that the structure was only a temporary one and that they planned to replace it with a double-track iron bridge within a few years. They explained that when they had attempted to buy the superstructure framing, the war between the North and South was still raging and the cost of iron was prohibitive. Their strategy, they said, was to build and live for a few years with a single-track wood bridge until the price of iron declined, and then they would put things right.[13]

FIGURE 3 The original Livingston Avenue Bridge at Albany, built mostly of wood, is shown here as it appeared in 1866, when this drawing (prepared from a photograph) was published in *Harper's Weekly*. Image provided courtesy of HarpWeek, LLC.

The population's disappointment wasn't limited to the choice of materials for the superstructure or to its single-track configuration. The community didn't much like the bridge's form either. Its long "S" shape stretched almost 2,000 feet across the river, incorporating trusses of varying heights, and it was flanked by about 1,500 feet of trestle work on the Albany side and about 800 feet of similar construction on the other side. The consensus was that the bridge was just plain ugly. Even the railroad station the bridge company proposed for the Albany side failed to please most people. It was a modest wood-frame structure that compared poorly with the more substantial building that the Central had been operating for more than twenty years less than a mile south of the new bridge and conveniently located near the popular Delavan Hotel. Even the location of the new bridge itself was the basis for some vigorous grousing. Its critics argued that the bridge should have been built farther south, where it would have been closer to the center of the city's activity.

Public opinion notwithstanding, work on the superstructure proceeded apace. By the closing days of 1865 the first Albany railroad bridge was substantially complete, and arrangements for opening it to rail traffic early in 1866 were underway. Even though it wasn't what most people had expected, the citizens of Albany and Greenbush were beginning to get used to it, and by then, after all the years they had waited for a bridge, they decided to make its arrival the subject of a rousing welcome. A fortuitous decision by the state legislature to make George Washington's birthday a legal holiday (for the first time) suggested 22 February 1866 as a good date for the opening ceremonies, and it proved to be just that. The day was cold but sparklingly clear, and a huge holiday crowd turned out along the waterfront to hear two military bands play and countless public speakers extol the virtues of the new bridge and its builders. With the Hudson solidly frozen, denying ferries the ability to cross, there was special appreciation of a New York Central locomotive with a four-car train in tow carrying officers and directors of the participating railroads eastward across the new bridge from Albany. Once across, the Central's locomotive was disconnected and replaced by a Hudson River Railroad locomotive, which returned the train to the Albany side with additional passengers from the Western Railroad (by then called the Boston and Albany) and from the Hudson River Railroad and the Harlem Railroad, dramatizing the symbolic coupling of the two sides of the river. Not more than a few minutes later, the sound of a train whistle on the Greenbush side signaled the approach of still another locomotive, this one traveling west at high speed along the easterly approach to the bridge and bound for Albany, Schenectady, Syracuse, and points west with eight loaded freight cars. As it crossed, the engineer waved, the bands played, and the crowd applauded. A new day had dawned for Albany.[14]

A little more than two years later, in 1868, the New York State Legislature made honest men of the railroad directors. It passed a bill in April of that year authorizing the Hudson River Bridge Company to construct a *second bridge* across the Hudson at Albany "in lieu of its present bridge." As soon as the new bridge was able to carry railroad traffic, the wooden bridge the company had completed and opened in 1866 was to be demolished and its debris removed from the river. The requirement for demolishing and removing the existing bridge was reinforced by a provision that required the company to post a bond of $600,000 to ensure that the removal would be complete and timely.[15]

By the time the second bridge was authorized, Erastus Corning had retired and been succeeded as president of the New York Central by Dean Richmond. Cornelius Vanderbilt, who had already made his first fortune in steamboats and was well along in the process of acquiring control of the entire New York Central system, had acquired a strong position in the Hudson River Railroad and through it was able to add his own considerable power to the efforts of advocates for a second bridge.

Momentum for proceeding with construction was even stronger than it had been for the first bridge, and things moved quickly this time. Within less than another year a change in the strategy of the bridge company had developed. It wasn't really going to demolish that first bridge, after all. Instead, it was planning to remove the first bridge's wood superstructure and replace it with a double-track, iron superstructure, and the work was going to be started just as soon as construction of the second bridge was complete. In May 1869 the legislature refined and strengthened its 1868 authorization to build the second bridge, this time defining as its western terminus a place near Maiden Lane on the Albany side and as its eastern terminus the old ferry slip at Greenbush (later part of Troy). The new act made no mention of demolishing the old bridge. It seems that among the members of the legislature, the management of the bridge company, and the new and powerful directors of the railroad companies, an understanding had been reached: getting rid of the first bridge just didn't make much sense anymore. The demand for rail transportation across the Hudson at

Albany had exceeded all expectations, and to serve it, the community really needed two double-track, iron bridges, not just one.[16]

Interest in conversion of the wooden bridge to iron waned briefly, as attention shifted to the urgently needed construction of the second bridge. The new span would be called the Maiden Lane Bridge, after the street that marked its Albany terminus, and by June 1870 foundation work was underway. Its piers and abutments, like those of the Livingston Avenue Bridge, would be built of stone and supported by timber piles; but this time the bridge company would contract out construction of the foundations to Charles Newman of Hudson, New York, who would bring the stone from quarries in Schoharie and Tribes Hill, a couple of New York State towns within about fifty miles of the site.[17]

Selecting a firm to design, fabricate, and erect the first iron bridge to cross the Hudson was no easy task. Europeans had been building iron bridges for a while, including a few spectacular ones, and some railroads (including the New York Central) had begun to build their share in America, too, but there certainly was no abundance of bridge builders who could offer substantial experience in the design and construction of iron bridges on the scale of what was planned for Albany.

One exception was the Clarke, Reeves Company of Phoenixville, Pennsylvania (for a few years before 1870 called Kellogg, Clarke and for many years after 1884 called the Phoenix Bridge Company). The firm was rich in the financial capital, talent, and bridge-related patents that it had inherited from or shared with its powerful parent, the giant Phoenix Iron Works. That very large and prestigious firm had since 1790 been producing and fabricating a wide variety of iron products that reflected the ebb and flow of the company's ownership and of the changing American culture itself. Unlike later firms that would concentrate exclusively on designing or fabricating or erecting ironwork, Phoenix Iron (and a small number of similar firms of the time) did the designing and fabricating in its own shops or in the shops of wholly owned subsidiaries, and then went out and erected the work with its own forces.

Phoenix's process was the ultimate in vertical integration, starting with unprocessed ore right from the mine. Then the company created the required iron "pigs" in its own foundries, and from those it shaped and fabricated its own iron members. In its earliest days, the company had produced ordinary nails in vast quantities, but by mid-century, as the railroads began to grow and spread, it had moved into large-scale production of the iron rails that carried the trains. During the Civil War it had developed and produced the Griffen Gun, an iron cannon that's credited with a decisive role in the Union Army's victory at Gettysburg, and it later established and patented more than a few designs for a wide range of iron applications that were natural by-products of the work it was doing. Its widely used "built up" column assembly ("the Phoenix Column") was immensely popular during the nineteenth-century proliferation of new buildings.

Not long after the Civil War, Phoenix Iron established its own bridge company, generating a reliable source of demand for its structural products and keeping the bridge work conveniently under its own management and control. Later, in 1884, Clarke, Reeves would reorganize as Phoenix Bridge Company, the Phoenix Iron subsidiary that would be a national force in American bridge building until late in the twentieth century.[18]

As its volume of work increased over the years, Phoenix Iron and its subsidiaries became more comfortable about working with outside firms and engineers. One frequent joint venturer was Squire Whipple ("Squire" was his name, not his title), a pioneer in early bridge design, and it was in fact a Phoenix-Whipple patented truss that would be selected by Clarke, Reeves as the principal structural element for its Maiden Lane Bridge. Whipple's truss used mostly standardized members, for which most of the engineering work had already been done and which could be fabricated and delivered to the work site in relatively short time. Clarke, Reeves brochures boasted that the company's work was so well prepared in its shops that no expert mechanics were required on site and that a project could indeed be built "without fitting, filing or chipping," a compelling incentive to a customer like the Hudson River Bridge Company that needed to see its bridge ready for traffic as soon as possible. At the peak of the fabrication period for this second Albany bridge, Phoenix Iron is said to have been employing (for all the work going on in its shops) as many as 1,500 workers.[19]

The physical profile of the second bridge would be generally similar to that of its upstream wooden neighbor, but of course its elements would be made (mostly) of wrought iron. For about two-thirds of its length, the bridge would have 28-foot-high, riveted, wrought-iron trusses that would include top and bottom chords formed of built-up plate and angle sections. Shallower wrought-iron trusses ("lattice deck girders") would be used for the shorter spans nearer the shorelines (away from the ship channel), where supporting piers could be more closely spaced. The total length of the new bridge would be almost 2,500 feet, including the trestle sections that would be built out over adjacent land areas to bring the track level down to grade in Albany on the west and in Greenbush on the east.

As finally built, the Maiden Lane Bridge, which would for a while be called the Middle Bridge, was a bit shorter than its upstream neighbor, the Upper Bridge. It would provide 30 feet of clearance above mean low water (only 25 feet at mean high water), and its huge draw section, 40 feet longer than the big draw section of the Upper Bridge, would allow well over 100 feet of clear passageway in both directions. After a good deal of attention and some controversy, it was decided that the swing sections would be opened and closed manually, but there's unconfirmed evidence that a 10-horsepower engine was added later.[20]

Even the original complaints about the location of the first bridge and criticism of its railroad station would be addressed by the designers of the

FIGURE 4 The Maiden Lane Bridge at Albany is shown as it appeared in 1917, with the then new Albany Pier in the foreground. Maiden Lane Bridge and New Pier, 1917, gelatin silver print, ht. 7¼ in. × w. 9½ in. Albany Institute of History & Art Library, V44.

second bridge. The new bridge's Maiden Lane landing was established at what was then pretty much the ideal location for good access to Albany's principal activities, and an attractive new depot would be built nearby, conveniently close to the celebrated Delavan Hotel, where earlier prohibitions against the serving of alcohol had only recently been relaxed. Late in December 1871 the first fully loaded train made its way across the newly opened Maiden Lane Bridge, its narrow, wooden neighbor to the north still in full view.

That view of the controversial wooden bridge wouldn't be seen for long, as the bridge company had already entered into a contract for designing and erecting a wrought-iron replacement. For that project, it turned (somewhat surprisingly) to a man named Thomas Leighton, a clearly capable bridge builder, but one whose experience with iron bridges before 1871 had been limited to the construction of a single, very short, single-swing bridge for carrying horse cars and pedestrians (but not locomotives) across the Erie Canal in Rochester. Leighton, born a little over fifty years earlier near Augusta, Maine, was a product of the system that bred most of the nineteenth-century American bridge builders: childhood on a farm, limited formal education, and then an apprenticeship in carpentry. Early along, he had found his way into building wooden bridges, some of them for the expanding railroad system of Panama and some for the railroads of the northeastern United States, including several for the New York Central. By 1854, his work in western New York had taken him to Rochester, where he had settled, built a shop, and raised a family. The second Livingston Avenue Bridge would be the first of many iron railroad bridge projects that he would undertake during his sixty-seven-year lifetime.[21]

Because the piers had been sized for two sets of tracks in the original construction, Leighton had little foundation work to do beyond making minor alterations to adapt the existing piers for the base configurations of the iron trusses. The periodicals of the day paid extraordinarily little attention to the project. The relatively complex tasks of removing and disposing of the old wood structure in a way that would not disrupt the activities of the harbor and then erecting the new wrought-iron structure in its place do not seem to have been described anywhere. An engraving of the finished bridge that appeared in a later brochure of the Leighton Company shows an iron structure almost identical to the original wooden bridge, but with iron members more slender than the wood members they replaced and the original piers looking about like they always had. As to the actual schedule of the work, all that's known for certain is that construction started sometime during or after the early months of 1872 and that the bridge was completed by sometime during or before the late months of 1875, in time to be shown with its identifying double track in the 1876 edition of *The City Atlas of Albany*.

With these two bridges in place, the problem of traveling between Albany and Greenbush appeared to most people to have been solved. But not everyone agreed. As some local activists began to point out as early as 1872, these were strictly railroad bridges, lacking features that would make them useful to the people who lived in Albany and Greenbush. Both bridges had sidewalks, they admitted, but walking across on a blustery, winter day next to a speeding locomotive spewing sparks was an unattractive prospect. What these citizens wanted was a bridge that would make pedestrian traffic comfortable and accommodate the passage of horse-drawn carriages and wagons, still the vehicles of choice for transporting passengers and local freight in the fourth quarter of the nineteenth century. A railroad on a proposed third bridge would be acceptable too, they said, provided that it traveled on a separate level. There was speculation that the Boston, Hoosac and Western Railroad might want to use a proposed third bridge for access to Albany, but a firm proposal from that railroad or any other remained elusive.

A group composed of a few public-spirited citizens and an entrepreneur or two, and headed by a lawyer named C. Adams Stevens, petitioned the legislature for permission to build such a third bridge, and in 1872 their request was granted. What the legislature authorized was a privately underwritten, two-level, wrought-iron toll bridge, with carriageways and pedestrian walks on the lower level and two sets of tracks on the upper level. The legislature appointed seven citizens as commissioners and assigned them the task of fixing an exact location for the bridge.[22]

This third Albany bridge was destined for a long and difficult course, and establishing a location for it would prove to be only the first (and not the toughest) of the many problems that delayed it. One of the commissioners died soon after his appointment, and another resigned. The bridge plan was popular locally, but opposition to its construction had survived with enough political power to delay the process of fixing an exact location until substitute commissioners could be identified and appointed. Troy's role and influence among the opposition had weakened by this time, but the steamship interests had taken up the slack, and some of their arguments against the bridge did appear to have merit. Much of the work of the shipping companies involved the towing of long strings of heavily loaded barges and small boats, sometimes as many as forty at a time. The possibility that an unanticipated gust of wind or a change in the direction of stream flow might foul one of these long tows made the steamship people especially wary about the width of the passageway through the draw section. In addition, some of the companies were accustomed to loading and unloading freight along the Albany shoreline, and they complained that a bridge south of Maiden Lane would force them to accept an undesirable alternative loading area.

The arguments continued for a few years, during which ideas for addressing the complaints were developed and solutions for some of them were found. A proposal to allow the builders to acquire the publicly owned property from which the ferries departed and arrived, as locations for the new bridge landings, led to accusations of corruption and favoritism but gained some acceptance anyway. The press and a sizable portion of the public continued to support the bridge idea, despite the opposition of some commercial groups that saw a threat to their interests. In 1875 a subcommittee of the Committee on Commerce and Navigation of the New York State Assembly, a group in which the steamship interests and others opposed to the bridge appeared to be disproportionately represented, took testimony about the status of the bridge plan. After exploring the reasons for the delays, as well as what appeared to it to be an insufficient amount of capital and a few unsuitable members among the bridge company's managers, the committee recommended that the charter of the new bridge company be revoked. It was a terrible blow to the bridge advocates, but they persisted and were able to discourage the full legislature from accepting the recommendation. Little by little, they moved forward with design work, but progress was halting.[23]

Early in 1880, eight contentious years after the legislature had authorized construction of the third bridge, the group that had shepherded the bridge through the legislature and survived the onslaughts of its opponents decided to call it quits. They sold their charter to a new (less controversial) group, headed by a couple of substantial and well-respected local citizens: Jose F. de Navarro, an investor of means who had earlier provided the funding for what would become the Ingersoll Rand Company, and A. Bleeker Banks, an Albany publisher who had become mayor of Albany in 1876. Disputes over the precise location of the bridge having by then been settled in favor of using the old ferry slip just a few hundred yards south of the

Maiden Lane Bridge, the purchasers moved quickly to acquire the ferry properties from the city and then to begin the work.

Some engineering work had already been done by then, although it wasn't yet visible to the public. The Clarke, Reeves Company, soon to be renamed the Phoenix Bridge Company, had done some preliminary engineering work when the idea of a third bridge had first been floated, and it was able to dust off those early drawings and move ahead quickly. The contract for the stone piers went to a Philadelphia masonry contractor, and the project was at last able to get off to a meaningful start. The bridge's superstructure would be shorter than that of either of its predecessors, because the river at the selected location was only 900 feet wide, so there would be fewer piers to build and fewer iron trusses to fabricate and erect. Phoenix Iron is said to have operated its shop twenty-four hours a day, employing

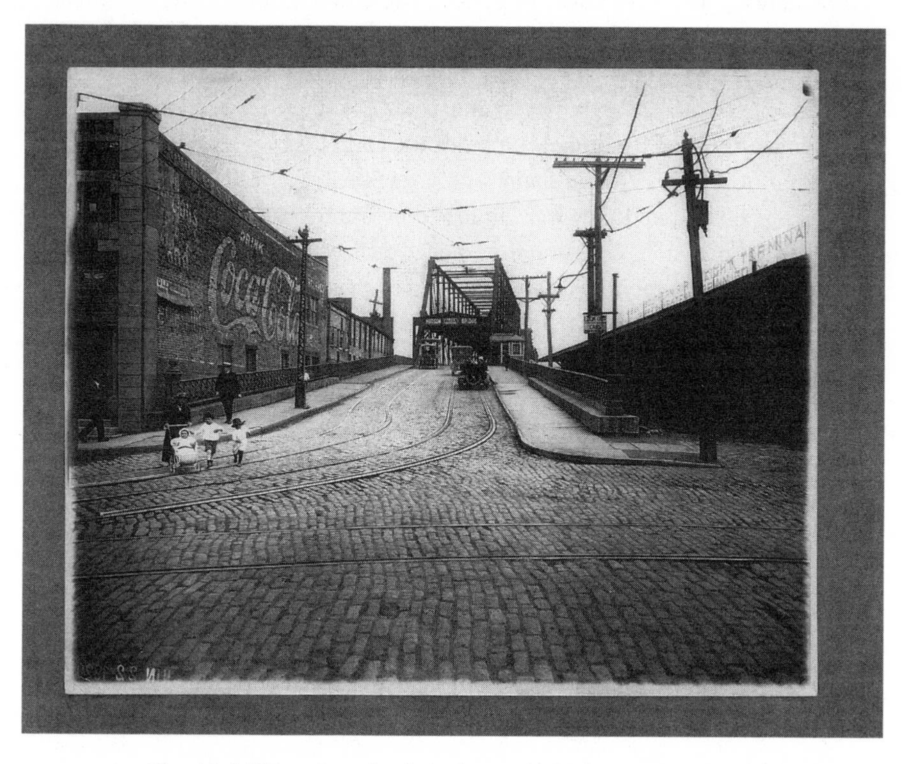

FIGURE 5 The third Albany-Greenbush Bridge is shown here as it appeared during the early 1930s. Note that its upper level, built to accommodate rail travel, has not been extended because no railroad ever used the bridge. Photograph courtesy of the Albany Public Library.

350 men for this project alone, to expedite fabrication and delivery of the ironwork to the bridge erectors. The 45-foot-deep trusses that were required to carry a two-level span required some of the heaviest members that Phoenix Iron had ever fabricated, and the total cost of the third crossing would—like the cost of the longer Maiden Lane Bridge to its north—prove to be about a million dollars. The huge draw section, said to be the largest ever built, would accommodate all the demands of the ship owners. At more than 400 feet wide, it would offer two passageways of a little less than 200 feet each, after allowing for the width of the center pier. The movable trusses of the draw section, about 50 feet high, would weigh about 250 tons each and would require a 35-horsepower engine to move them. The whole draw section was a colossal undertaking, widely noted around the country and studied by engineers and students for years.

All those challenges notwithstanding, the Lower Bridge, as it continued to be most popularly called, was completed and made available for carriages and pedestrians within only a little more than eight months of the start of fieldwork, a spectacular achievement. On 24 January 1882 twelve sleighs carrying officers of the bridge company, the contractors, the engineers, and a group of officials from Albany and Greenbush were drawn across the bridge by horses, and they were followed by hundreds of pedestrians celebrating the occasion. Crossing the bridge was free on opening day, but the collectors were at their posts at first light the next morning, collecting tolls that had been established at levels equal to those of the ferry that was being displaced. There was still no sign of a railroad for the new bridge when it was opened, and there was little evidence that there would ever be one.[24] Some years later, a trolley system was added on the lower level, but apparently no railroad company ever found the potential rewards of using the bridge sufficient to justify the cost of relocating its system and building the trestles needed to elevate tracks to the bridge's upper level.

The three railroad bridges provided well for connecting the east and west banks of the Hudson at Albany until about the turn of the twentieth century, when the wrought-iron superstructures of all three were replaced by steel structures. Only the Upper Bridge, now usually called the Livingston Avenue Bridge, survives for limited railroad service in the twenty-first century.

CHAPTER 4

The Last of the Railroad Bridges

Poughkeepsie and Castleton

News of all that bridge-building activity up at Albany and beyond certainly wasn't lost on the cities and towns of the central and lower Hudson Valley, where populations and the commerce they generated were growing apace.

One of those downstate cities was New York. By 1880 it would have a population of 1.7 million—up from 1 million only twenty years earlier and destined to double by 1900—and the city certainly didn't need reports of new upstate bridges to be reminded that it needed to find better ways of getting across the river. Despite a continuing flow of new ideas, it still relied on ferries that were vulnerable to ice, wind, fog, and a whole variety of mechanical and human weaknesses. Nevertheless, it would be decades before a bridge would connect the big city with New Jersey.

New York certainly wasn't the only place in the lower Hudson Valley where people wanted and needed to cross the river or where they saw a new bridge as something that might enhance their prosperity. As early as 1841, when the Erie Railroad had begun laying track westward from the village of Piermont, on the west bank of the lower Hudson, there had been brief hopes of building a Hudson River railroad bridge there. Piermont was about thirty miles north of the center of things in New York, and nearby Tarrytown and Nyack were a couple of river towns that had already established themselves as the east and west ends of a good crossing place. But when the Erie failed to show any real interest in crossing the river there and then established its own ferry service from Piermont to Manhattan, any justification for a bridge near Piermont (if there had ever really been any) disappeared. It would be another 114 years before the Tappan Zee Bridge would be built there.

Not much farther up the Hudson, where the river begins to narrow and the distances between ferry crossings get a little longer, serious talk about building a bridge just north of Peekskill was beginning to be heard around 1868. The rich and well-connected upstate railroad figure Erastus Corning, together with a few of his influential colleagues, was convinced that a railroad could profitably be brought east to Peekskill from the coal fields of Pennsylvania, carried across the Hudson by bridge, and then continued east to the thriving cities and factories of New England.[1] He established a corporation called the Hudson Highland Bridge Company, which assembled a panel of engineers to explore the subject of a bridge just north of Peekskill. It was a fairly capable group, headed by General George McClellan, but for the most part not the first tier of nineteenth-century engineering. McClellan had been seen as a promising young engineer when he graduated (second in his class) from West Point in 1846, but as a military leader during the Civil War he had disappointed President Abraham Lincoln and been sidelined. By the time the Hudson Highland project came along, a series of detours had taken their toll on his army career, and he was probably looking for a job. Horatio Allen, another well-established engineer on the panel, would within another few years succeed to the presidency of the prestigious American Society of Civil Engineers. He had for years concentrated on the design and construction of locomotives, but it's doubtful that he had ever had much to do with bridges. Another former military officer, a Civil War general named Edward W. Serrell, was clearly the most promising civil engineer of the lot. He had concentrated on bridge design and related work for years before and during the war, and he became Hudson Highland's chief engineer.

The engineers apparently discharged their responsibilities well, approving the concept, encouraging the organizers to proceed, and identifying a workable crossing place between Fort Montgomery, an elevated site on the river's west shore, and Anthony's Nose, a correspondingly high place on its east shore that's said (by some) to have been named for a local river captain whose big nose it resembled. The location lies a few miles north of Peekskill and approximately along the alignment of the later Bear Mountain Bridge, which would be built there in 1924. At the time they selected the site, the engineers made a recommendation that suggests they were a good deal more sophisticated in contemporaneous thinking about bridges than their backgrounds might have suggested. They elected to dodge the usual thorny issue of obstruction to river navigation by proposing a suspension bridge, which could be built at that narrow reach of the river without building any piers in the water.[2] The suspension concept was well known in the early 1870s, but it was for the most part still being considered mainly for major bridges. It wasn't the comfortable, well-tested design that conservative engineers like these fellows might have been expected to propose for a modest railroad bridge near the small town of Peekskill.

There was more than enough enthusiasm for the Peekskill Bridge, but it was premature. About thirty miles north of its site, another group of bridge advocates was gathering support for building a bridge at Poughkeepsie instead, and there was little likelihood that both bridges would be built. The idea of a bridge at Poughkeepsie had a lot going for it, and it had been around for a while. As early as 1855, a man from Mississippi who styled himself a "bridger" had written to the *Poughkeepsie Eagle* suggesting such a bridge, proposing to design and build it himself, and even quoting a price of $1 million for the job. Letters and editorials endorsing his idea appeared from time to time after that, and by the late 1860s the idea had gained some committed support and influential supporters. By then Poughkeepsie had become a city of about 20,000, three times the size of Peekskill, and it was a thriving, dynamic commercial center about halfway between New York and Albany. Well located for a bridge, it was a community that had the energy, resources, and determination needed to get one built.[3]

The city was the prospering market center for the agricultural communities that surrounded it, and its stable and substantial industrial base included Adriance-Platt Mowers, which was selling its farm equipment all over the country, and the Vassar Brewery, whose especially well-regarded ale is said to have enabled the company to grow every year. Poughkeepsie in the 1860s even had a couple of colleges: Vassar Female College (later Vassar College), founded in 1861 by the proprietor of (and from the funds generated by) the Vassar Brewery; and the Eastman Business College, founded in 1859 by Harvey Eastman, a smart and energetic local booster who was a cousin of the Eastman of Eastman Kodak.

Harvey Eastman adopted the Poughkeepsie bridge cause as his own early along, and soon after that things began to happen. In 1871 he got himself elected mayor, a position from which he could effectively seek the formal approvals needed to build a bridge and arrange for its financing and construction. By the summer of 1872, he and Pomeroy P. Dickinson, a civil engineer who had useful experience on some of the region's railroads, had shepherded a bill through the state legislature that authorized the incorporation of the Poughkeepsie Bridge Company. Eastman made a surprising early decision to engage Horatio Allen as chief engineer, even though Allen had been serving in exactly the same capacity for the group sponsoring the rival bridge at Peekskill. Allen acquitted himself well by quickly establishing a structural concept for the Poughkeepsie bridge. Like almost everyone involved, he had been thinking in terms of a suspension bridge, and his idea was to adapt to the Poughkeepsie site the concept that had been proposed for the bridge near Peekskill. But the narrow river that the Hudson is at Peekskill widens at Poughkeepsie to almost 2,700 feet, too wide for a single suspended span in 1871.[4] Allen's concept at Poughkeepsie would require at least two piers in the water to shorten the suspended span, and he quickly realized that those piers could become controversial.

Once he moved from the conference room to the drafting room, Allen became aware (or remembered) that the legislative authorization had included a troubling but apparently unnoticed paragraph that specifically barred the placing of any piers at all in the water. Such a constraint, he realized, could mean the end of the Poughkeepsie bridge. Without intermediate supports, the bridge would have had to make the crossing in a single leap, something that was beyond the technology of the time. The fierce opposition of the shipping interests, the same ones that had resisted the bridge builders in Albany and Troy, was behind the restriction.

Eastman was disappointed, of course, but not discouraged, and later in 1872 he reacted with a couple of practical and effective strategies. One was to seek the opinion of an acknowledged expert as to just what kind of bridge Poughkeepsie should be building, and the second was to do whatever was necessary to revise the existing legislation.

For his expert, Eastman turned to an engineer named James Buchanan Eads, who was well along in supervising the construction of a multiple-arch, steel bridge that would cross the Mississippi at St. Louis on three soaring 500-foot arches. It was one of the boldest designs of its time, and Eads (whose middle name, Buchanan, was taken from his mother's cousin, the country's fifteenth president) was the gifted, self-taught engineer who had designed it.[5] Fifty-one years old when he met with Eastman in St. Louis about the Poughkeepsie bridge, Eads had already established himself as one of those nineteenth-century figures who were able to advance the technologies of the time with a blend of great intelligence and extraordinary mechanical talent. Raised along the Mississippi and knowledgeable about its idiosyncrasies, he had in his twenties invented and built a submersible diving bell in which he had been able to explore and record useful information about the river and to find and recover some of the valuable sunken treasure that had for years been accumulating along its bottom. His successes in salvage work made him famous in the region and very rich. During the Civil War, Lincoln had selected him to manage the design and construction of a fleet of iron-clad vessels that would influence the course of the war and become models for later warships. A few years after the end of the war, Eads would be selected to do what Charles Ellet, John Roebling, and a couple of other distinguished engineers had failed to do: design and build a bridge across the Mississippi from St. Louis to East St. Louis.

Harvey Eastman was a practical fellow with a healthy respect for the power of the shipping interests, and he reasoned that trying to accommodate their objections would probably be the best way of getting them to relax their opposition to a bridge. It wasn't just the piers that worried them, he learned; it was the temporary falsework, the scaffolding that needed to be built in the river (usually right in the middle of the shipping channel) to support the structural frame until it could support itself. At St. Louis, Eads had developed a technique that eliminated the need for such temporary support. He was building the separate halves of each arch simultaneously, supporting each unfinished half-arch from above, as a cantilever, by tying it back to a high, temporary tower he had built on the adjacent pier. Once those two halves of the arch met in mid-span and were connected to one another, the whole arch became self-supporting, and the temporary towers and cantilevered framing that had been holding them up could be removed.[6]

There's no clear evidence that Eads recommended his arch system to Eastman and Allen for Poughkeepsie, but neither did he endorse the suspension plan in the opinion he gave them. Instead, he recommended that they make it convincingly clear to the New York State Legislature that they must be allowed to build four piers in the river, two more than Allen had been advocating. That, Eads explained, would allow a configuration of supports that—when combined with a pier on each shore—would allow for crossing the river at Poughkeepsie with five very long truss spans of about 500 feet each. The piers would be far enough apart to satisfy the objections of the ship captains, and although the spans would be long enough to be challenging, they were not beyond the capabilities of the period. It was an eminently practical way of configuring the bridge, but of course it still did not solve the nagging problem of avoiding falsework in the river. Conventional 500-foot trusses on piers would have to be built in place, and they would require a forest of falsework to support them until they could support themselves. Whether Eads assumed that the designers of a bridge at Poughkeepsie would figure out for themselves that they could apply his cantilevering technique to a truss bridge even more easily than he had done with his arches is not known for certain, but later evidence suggests that he had just such an approach in mind in 1872.

Satisfied with what he had heard from Eads, Eastman moved as quickly as the processes of civil government would allow to induce the legislature to revise the bridge's charter.[7] Starting with an announcement that he had become a candidate for election to the state legislature, he followed with a promise that, if elected, he would shepherd the necessary bridge legislation through to ratification himself. His plan worked. Eastman was elected by a wide margin in 1872 and took his seat in Albany the following spring, ready to tackle the forces arrayed against the bridge. The opposition, led by the shipping companies and supported by Albany business interests who saw Poughkeepsie as a potential competitor, was well funded and formidable, and the principals aligned against the bridge were spoiling for a fight. But in the months that followed, a few previously unknown supporters emerged, and (with the help of some behind-the-scenes skullduggery that is said to have been worked by the Eastman group) the bridge advocates managed to obtain an amended charter that allowed four piers in the water and provided that construction of the new bridge must get underway by the end of 1873. It was time to start raising money and identifying the railroads that would use the bridge.

Whether it was Harvey Eastman's visit with James Eads in St. Louis that first brought him into contact with Andrew Carnegie isn't clear. Carnegie was the principal owner of Keystone Bridge Company, the firm building the Eads bridge, and Carnegie is known to have been a frequent visitor to the construction site. The extraordinary Carnegie, who had barely turned thirty-seven in 1872, had by then been retired for seven years from his position as superintendent of the Pittsburgh division of the Pennsylvania Railroad and was already well on his way to becoming one of the richest and most powerful men in the world.[8]

The investment that Carnegie identified as his favorite in those years was Keystone Bridge of Pittsburgh, a fabricator and erector of iron bridges.[9] The Keystone investment reflected Carnegie's confidence that the growth of the country was bound to require a vast expansion of its railroads, an industry he already understood well and in which he was heavily invested. He also recognized that such an expansion would rely on the construction of many miles of track and of a good many iron and steel bridges. By 1866 he had merged Keystone with a producer of iron shapes and plates to form one of the first of the vertically integrated iron manufacturing and fabrication companies in the country. In another few years that firm and a few others would become the core of his Carnegie Steel Company, the principal element at the turn of the century in the conglomerate that J. P. Morgan would call U.S. Steel.

In 1871, when Keystone was building the Eads bridge, Carnegie's close friends and colleagues at the Pennsylvania Railroad, which was itself a principal in the ownership of the bridge and about to become its most valuable customer, were monitoring progress at St. Louis with more than casual interest. Carnegie was a regular visitor to the work site, as his interests there were vital and broad. In addition to holding the majority position in Keystone, he was a key figure in the financing of the bridge and would soon be off to London to sell its bonds, on a commission basis, to the banking firm of J. S. Morgan Company, the parent of the later, U.S.-based J. P. Morgan Company.

Carnegie's energies and his entrepreneurial objectives knew few limits. Once he learned from Eastman that Poughkeepsie was planning to build a bridge that would be half again as long as the Eads bridge and likely to be going into construction just about when Keystone would be winding down its work in St. Louis, he recognized a potential contract for his company. And that wasn't all. At the time of Harvey Eastman's conversations with James Eads, Carnegie was a member of a committee that had been established by the president of the Pennsylvania Railroad to explore ways in which the Pennsylvania could extend its reach from the coal fields of Pennsylvania into the industrial centers of New England, the very idea that was at the heart of the plan to build a bridge at Poughkeepsie. The timing of all these events was wonderfully fortuitous, and if there was anyone better positioned than Carnegie to bring it all together, it's hard to imagine who that might have been.

Neither Eastman nor Carnegie wasted any time. An agreement was quickly reached that made Keystone the contractor for the bridge at Poughkeepsie. Keystone's chief engineer, Jacob Linville, headed up design and construction, with Pomeroy P. Dickinson looking over his shoulder for the

Poughkeepsie owners. Linville was already well established in the field, having designed several long-span truss bridges, including an especially noteworthy one across the Ohio River at Stuebenville, and he had supervised construction of the Eads bridge as Keystone's chief engineer.

In fact, Linville and Eads had been at loggerheads during much of the St. Louis construction, and at Poughkeepsie Linville quickly made it clear that he was no mere acolyte of Eads. He would design this new bridge to suit the four-pier plan that had been authorized, he declared, but it would be a conventional structure that employed very deep, parallel chord trusses. There would be no cantilevers in Linville's design for a bridge at Poughkeepsie. It's not clear how he was planning to deal with the inevitable opposition of the shipping crowd. Dickinson, who was at that time still associated with the locally based Poughkeepsie and Eastern Railroad, had a more or less passive role at this stage, looking after the owners' interests but leaving design decisions to Linville.

While issues of design were being explored, the economics of the bridge had to be addressed. To be viable, the new bridge would need a good railroad connection at each end to ensure enough traffic. At the eastern end, Dickinson's Poughkeepsie and Eastern Railroad had recently completed a line from Poughkeepsie to the Connecticut border, where it anticipated a connection that would take its trains east through Hartford and on to Boston. The situation at the western end of the bridge was less certain. There, nine miles of railroad would have to be built by someone before trains crossing at Poughkeepsie could connect with lines that would take them into the coal country of Pennsylvania. Once again, Andrew Carnegie's relationship with the Pennsylvania Railroad entered the planning, and he was able to secure assurances that the Pennsylvania would close that gap before construction of the bridge was finished.

Financing for building the bridge was of course a critical issue, and once again Carnegie's network of connections and his influence were invaluable. In 1873 he reported to Eastman that his friends J. Edgar Thomson, president of the Pennsylvania, and A. L. Dennis,[10] a wealthy banking member of Thomson's advisory committee, would be making a joint personal investment in the new bridge, committing to the purchase of $1.1 million worth of shares. The commitment was just what was needed to get design work and construction underway.

Everything was falling nicely into place, except for the still missing strategy for dealing with the shipping interests. In December 1873 Keystone made a well-publicized start on building one of the easily accessible shoreline piers. Sound rock could be reached there by conventional methods of excavation, without pile driving, dewatering, and the like, and just days before snow began to shut down construction in Poughkeepsie, a Keystone crew was able to put the eastern shore abutment in place. Dignitaries and local citizens gathered on 29 December 1873 to witness and celebrate the laying of a cornerstone, and the stage was set for the great bridge that would follow.

But it wasn't to be.[11] The sometimes frenetic and often unjustified expansion of railroads that had followed the end of the Civil War had by 1873 combined with excessive financial speculation to produce a badly overextended national economy. By fall of that year, it had all begun to unravel, eventually bankrupting about a quarter of the country's railroads and bringing on a depression that would last for years. The Pennsylvania Railroad would itself survive, but its directors lost their taste for new ventures, at least for a while, and they backed away from assurances that there would be good railroad connections on the west end when the bridge was finished. Then J. Edgar Thomson died in 1874, depriving the bridge company of its wealthiest patron. Thomson's estate declined to implement his commitment to purchase a large block of bridge stock. One by one, the underpinnings of Eastman's elaborately organized plan for building the bridge collapsed, and what had promised to be Keystone's splendid first days of construction in 1874 proved instead to be its quiet last days in Poughkeepsie. The issue of obstructing river traffic with falsework for the steel did not have to be faced. Andrew Carnegie headed back to Pittsburgh to build the first of his steel mills there, and the bridge at Poughkeepsie, for the time being at least, disappeared from public discourse.

Almost everyone in commerce (except the financially independent Andrew Carnegie) was immobilized by the Panic of 1873, so the bridge builders of Poughkeepsie didn't lose much ground to their struggling competitors down near Peekskill. Ironically, a pamphlet that appeared in that troubled year added a third proposal for a railroad bridge in the mid-Hudson area, a crossing that would be suspended far above the water's surface near Storm King Mountain, about seventeen miles north of the Poughkeepsie site.[12] The width of the river at that location was about the same as at Poughkeepsie, and although early publicity suggested a suspension bridge there, in later versions cooler heads would prevail and a multi-span steel truss bridge, not unlike the one proposed for Poughkeepsie, would emerge as a model. A later pamphlet implied that the Storm King bridge had strong railroad support on both sides of the river, but it was too early to tell whether there was really anything to that claim. As yet, the Storm King backers had no charter from the state and didn't appear to be seeking one, so the Poughkeepsie crowd wasn't worrying much.

About fifteen miles downstream from the now abandoned Poughkeepsie bridge site, ferry operators at the bustling river port of Newburgh continued to have the cross-river freight business pretty much to themselves. The closest railroad crossing was still more than ninety miles north, in Albany.

The failure of the Poughkeepsie plan was a terrible blow to advocates of the bridge, but a core group rallied around Eastman in an effort to fund the bridge all over again. A campaign to sell stock was at the heart of the new plan, and some success was achieved by inviting subscriptions that did not have to be paid until at least a million dollars' worth of shares had been sold. Even under that arrangement, the effort to raise money for a railroad venture during a period of economic depression that had its roots in railroad expansion was understandably difficult. By 1875 Eastman and his staff had begun to consider a new source: the bridge-building contractors themselves. Talking to them one by one, Eastman identified a few who showed interest in a strategy that would combine their traditional roles as engineers and builders with the new role of investor. By the end of the year he was approaching an agreement with one of the largest and best-regarded of the firms, the American Bridge Company of Chicago. Having just finished work on a challenging lift bridge across the Missouri River at Boonville, American Bridge seemed a natural candidate, probably the most promising of the lot: a capable and wealthy company in serious need of a new job. American Bridge's proposal and its cash had the effect of meeting the requirement that one million dollars had to be raised, and once again Eastman had a viable (if still slightly shaky) basis for moving the project forward.[13] Once the proposal was accepted, early in 1876, the previously conditional subscriptions were converted to cash, and the contractor moved quickly. But by the time American Bridge had mobilized its forces and had adapted the previously developed drawings to suit its own practices, the summer of 1876 was turning to fall.[14]

W. G. Coolidge, a seasoned civil engineer who had managed the work at Boonville for American Bridge, took charge of on-site operations, while the reliable Pomeroy P. Dickinson again looked after the interests of the owners. Just how this bridge would be built was going to be decided this time by Coolidge, and in late 1876 his attention was focused on the challenge of building the four massive river piers the legislature had allowed. All four were designed to rest on bedrock that in some cases was as much as 140 feet below high water.

Like Keystone's Linville, Coolidge had no taste for experimenting with cantilevers and was content to let the bridge owners face the anger of the shipping interests and others. And he was not enthusiastic about the pneumatic caisson approach that Eads had used about five years earlier in St. Louis and that Roebling was using in Brooklyn. For his deep piers, Coolidge favored the fairly primitive but well-tested idea of building and sinking large, heavily reinforced, watertight timber cribs through which dredging buckets could remove the muddy clay that covered the bedrock, and which would ultimately become the concrete-filled foundations of the bridge itself. These were the timber forerunners of what would in a later period be called open-dredged caissons.

Construction of the first of the big cribs began in November 1876, on the very docks on which some of the still young country's first warships had been fitted out almost a hundred years earlier, and a second crib was started a month or two later. The first crib measured about 50 feet by 100 feet, a giant structure that covered an area almost as large as a couple of standard tennis courts, and its sides were carried to a height of about 30 feet, with provision for extending them much higher once the crib was placed in its final position in the river. Inside, the space was subdivided by thick timber bulkheads that defined individual, vertical shafts, some of which would be filled with ballast to cause the floating crib to sink when it had been towed to its final location, but most of which would be left open to allow the dredging buckets of barge-mounted equipment to excavate and remove material from the river bottom. By February 1877, some river ice in the salty Hudson notwithstanding, the first crib had been floated to a point about 500 feet from the west shore, where the weighting chambers were filled and the whole massive structure was allowed to force its iron-shod cutting edges down into the muddy bottom of the river.

By summer, the process had demonstrated its elegant effectiveness. Excavation for the first foundation had been completed, the first crib's walls had been built up to within about 15 feet of high water, and the structure had been filled with concrete. A second foundation had been brought almost to the same level, and the crib for a third was under construction at the dock. The next step was to build above each of the concrete foundations a big stone pier that would directly support the bridge framing above it. To allow for men to do the required stonework, much of which would have to be built (and would remain permanently) below the water's surface, temporary watertight enclosures (cofferdams) would have to be built and anchored to the top surface of the concrete foundations. By fall those enclosures had been built on two completed concrete foundations, but the pressure of the deep water they had been designed to resist proved too much for one of them, and it failed. The big cofferdam itself was destroyed, and it damaged the foundation to which it had been secured, leaving American Bridge Company with the formidable (and expensive) tasks of rebuilding the cofferdam and restoring the foundation.[15]

The company made a genuine effort to restore the damaged pier and continue the work, but it proved to be a huge struggle. By the time the winter weather of 1877–1878 brought work to a stop, American had restored the damaged concrete and rebuilt the cofferdam, and workers had even managed to bring the stonework to levels above high water. But the company had been mortally wounded. The combination of this failure and other depression-related reverses had been too much, and it was forced into bankruptcy early in 1878. Efforts to bring in other contractors to finish the work were unavailing, and by mid-1878 the second attempt to build the bridge had ended in failure. Later in the year, Harvey Eastman died of tuberculosis at the age of forty-eight. The prospect of ever seeing a completed bridge at Poughkeepsie seemed more remote than ever.

Bridge loyalists may have taken some comfort from knowing that their competitors at Anthony's Nose and Storm King weren't doing much better, although neither of those rivals had yet quit. Nothing had been built at Anthony's Nose, but General Serrell, one of the consulting engineers, had taken over the project when Erastus Corning died, and he had apparently brought some new vigor to the program. At Storm King, prospects for a bridge seemed less good, although there were reports that some railroads had made meaningful commitments and that more were coming along every day.

The city of Poughkeepsie itself would need a while to recover from the hard times that had started with the 1873 financial panic. The city and some of its local investors had lost a good deal of money when the Poughkeepsie and Eastern Railroad failed, and it wasn't long before new laws were enacted to deny towns and cities the legal right to invest in railroads. Even the population of Poughkeepsie itself, which had been growing steadily (along with the tax revenues the city collected and the general level of prosperity), failed to increase at all between 1870 and 1880. With population growth stalled, railroads failing, and once-solid businesses entering bankruptcy, it wasn't surprising that investors for a third attempt to build a railroad bridge across the Hudson were hard to find.

But the country was still young and resilient. By the early 1880s, the economy had begun to recover, and plans for several important new bridges in the region were underway. In 1882 the third bridge at Albany was completed, and it was followed in 1883 by the great Brooklyn Bridge. A few miles south of Poughkeepsie, the Newburgh ferries were busier than ever, carrying railroad cars loaded with Pennsylvania coal and dramatizing the justification for a bridge that would provide a shorter, weather-independent route between Pennsylvania and New England. Not surprisingly, there was a renewal of interest in building a railroad bridge at Poughkeepsie.

A wealthy Philadelphia utility executive named William W. Gibbs became interested enough in the Poughkeepsie bridge project to acquire control of its severely discounted stock shares, and he was able to round up enough additional investors to provide a starting basis for a third try at building a bridge.[16] The company he established arranged a contract with the newly formed Manhattan Bridge Company to manage the whole process as a general contractor. Manhattan was headed by another Philadelphia investor/businessman, and its directors and shareholders included a preponderance of wealthy Philadelphians, a Frick or two from Pittsburgh, Henry Seixas in New Orleans, and a few other well-known nineteenth-century financiers. The new crowd moved quickly to arrange financing for the bridge, and by late 1885 the stage had been set for a third try. The next step was to identify and engage a capable engineering and construction company to design and build the bridge.

Early in 1886 there wasn't much about building a bridge across the Hudson that would be new to a distinguished fifty-nine-year-old Massachusetts-born civil engineer named Thomas Curtis Clarke. Before his fortieth birthday, Clarke had designed at least one major bridge across the Mississippi and more than a few other major bridges and structures in various parts of the Midwest. In 1868 he and a colleague named Charles Kellogg had joined forces to establish Kellogg, Clarke and Company as the bridge-building division of Phoenix Iron Works, the already seventy-eight-year-old Pennsylvania company that had grown from modest beginnings as a nail manufacturer into one of the country's largest producers of railroad track, with 1,500 employees and few peers in the business.

Clarke was no stranger to the Hudson. He was the Clarke of the Clarke, Reeves firm that had built a couple of the bridges at Albany. Between 1870 and 1872, he and his partner Kellogg had designed and built the Maiden Lane Bridge there. When Kellogg left to pursue other interests, Clarke teamed up with Samuel Reeves, another principal of Phoenix Iron Works, to form a successor bridge-building firm, which came back to Albany to build the double-tiered railroad bridge just south of the Maiden Lane. When that bridge opened in 1882, there was probably no engineer in the country with more experience in bridging the Hudson than Thomas Clarke. But it wasn't just Clarke's experience and his success in building a couple of major bridges across the Hudson that made him as interesting as he was to the sponsors of the Poughkeepsie bridge. Rather, they were attracted by the people he had teamed up with.

A couple of years after the work at Albany had wound down, Clarke ended his sixteen-year association with Phoenix to become one of five principals in the newly organized Union Bridge Company of New York, a firm established to merge the old Kellogg and Maurice Bridge Company of Athens, Pennsylvania (a company that had no connection with the Kellogg of Clarke's earlier years), with the equally large and important Central Bridge Works of Buffalo, New York. The Kellogg and Maurice firm had been owned by Charles Kellogg and C. S. Maurice, but Kellogg sold his interest to Charles Macdonald, an especially distinguished Rensselaer-educated engineer with a long list of important bridges to his own credit. Maurice would remain as a principal in the Athens firm, managing the sprawling fabricating shops the company maintained there, which were said to be the country's largest, and Macdonald would head up design and engineering. Central Bridge Works, the other half of the new Union Bridge Company, was owned by a retired general named George Field, who had responsibility for its field operations, and by a civil engineer named Edmund Hayes, who ran the big fabricating shop the company maintained in Buffalo.

The four owners of the two merging firms, together with Thomas Clarke, clearly represented exactly the broad and deep capability that was being sought for the third try at Poughkeepsie. With Clarke as its elder statesman, Macdonald as its principal designer, Maurice and Hayes as managers of its shops, and Field as its manager of operations, Union was a large presence in bridge construction, second to none. It was a presence well appreciated by the directors of the Manhattan Bridge Company, which awarded Union Bridge a contract early in 1886 to design and build a bridge across the Hudson at Poughkeepsie.[17]

That's when a splendid bonus surfaced. Only a few years earlier, Edmund Hayes and George Field, at that time doing business in Buffalo as Central Bridge Works, had completed construction of a spectacular cantilevered bridge across the Niagara River that would have profound implications for the design of the Poughkeepsie bridge. On the Niagara job, they had to cross an intimidating gorge that was about as nasty a place as could be found for building the timber falsework needed for a conventional span, and local geology and other factors argued against a suspension bridge. What the distinguished engineer C. C. Schneider designed instead, and what the Central Bridge Works built, was the second of the first two American truss bridges to rely on the cantilever principle that Eads had used for erecting his arches on the Mississippi. At last, the possibility of maintaining clearance for ship owners, even during construction, appeared to have a chance.[18]

The cantilever principle for bridges had begun to gain wider acceptance in Europe, where the work of a German engineer named Heinrich Gerber was becoming popular. Gerber's bridges were crossing deep and terrifying gorges in mountainous country without the use of falsework. Gerber was cantilevering long sections from adjacent towers and then filling the gap between their unsupported ends with conventional trusses, obviating the use of falsework.[19] Neither Keystone nor American Bridge, when each made its failed attempt to cross the Hudson, had dared to try the cantilever ideas of Eads and Gerber as ways of keeping the river's shipping channels open, but now it appeared that Union Bridge would bring to Poughkeepsie the experience that would make such an approach feasible.

By late 1886, the company settled into designing and building the new bridge for Poughkeepsie. The engineers decided early along that they would stay with the approximately 500-foot-long river spans that had been the basis of the two preceding efforts, preserving—for the most part—the foundation work that American Bridge had already completed on two of the four river piers.[20] And it wasn't long before they decided that their bridge would definitely be built as a cantilevered structure. Two of the five over-water spans would still require falsework in the river, but they were not located over the shipping channel. The other three over-water spans, including the one over the shipping channel, would be permanently free of falsework and likely to accommodate even the harshest demands of the ship captains.

Not surprisingly, Union also brought along a few other design ideas, one of which required that virtually the entire superstructure be built of mild steel. The 1870s and 1880s had seen real improvement in the production of steel (and a reduction in its once prohibitive cost), and although Union would still use some wrought iron in the less critical elements of the structure, like the approach viaducts, Clarke was personally committed to the superiority of the new mild steel for the principal elements of the bridge.[21]

To manage construction of the bridge, Union brought in a talented Irish-born thirty-two-year-old engineer named John F. O'Rourke, who had graduated from the Cooper Union only ten years earlier and had been working in bridge construction ever since. Soon after finishing work on the Poughkeepsie bridge, O'Rourke, whose brogue and flowing moustache enhanced his personal charm, would marry the beautiful daughter of a wealthy Poughkeepsie manufacturer and move on to a spectacular career in New York, where his own construction firm would become one of the city's largest and most successful. Early along, as the plan for erecting the Poughkeepsie bridge took shape, he and his colleagues at Union Bridge set about dividing up and subcontracting to others virtually all the work required except the fabrication of the structural steel and iron work, which was the specialty of Union Bridge. The extensive subcontracting of work as long ago as 1886 weakens the widely held modern notion that subcontracting of construction work is a recent innovation. At Poughkeepsie, under Union Bridge's supervision, subcontractors would do all the dredging, pile driving for the falsework, construction of falsework, concrete foundations, and masonry piers for the bridge, and even erection of the structural steel itself, once it had been fabricated and delivered to the site by Union Bridge. And subcontractors would paint the steel and lay the railroad tracks, too.[22]

Once the contract to design and build the bridge had been signed and the design work had been started, it didn't take Union Bridge long to build momentum in New York. Fabrication of steel in its shops in Athens, Pennsylvania, and Buffalo couldn't get underway for some months, of course, while engineering work was being done and shop details were being prepared, but by September 1886 O'Rourke had set up his base of operations in Poughkeepsie and begun mobilizing the personnel and materials that would be required. Within another month, survey parties would be establishing baselines from which the precise locations of anchorages and piers would be determined. The docks along the Poughkeepsie waterfront began filling with heavy hemlock and oak timbers for building the underwater cribs and with lighter timber for building the falsework. Long lengths of yellow pine for piling to support the falsework and a wide variety of construction equipment and supplies that would be needed began arriving by barge and would soon pretty much cover the rest of the waterfront.

Construction of the piers came first, and here Union was for the most part constrained to follow a course that had been set ten years earlier by American Bridge, which had put two of them in place and had started the timber crib for a third before giving up. As in the earlier effort, the strategy was to build huge floating timber cribs, tow them out to their intended locations, and then lower the buckets of barge-mounted dredging equipment down through the cribs to excavate material from underneath them. Little by little, the excavated material would be replaced by concrete, and the big foundations for the bridge would take shape.

Under O'Rourke's supervision, the foundation work proceeded more or less routinely through the early spring of 1887, with little more than the usual small disasters that accompany most marine construction. But then things took a discouraging turn. One of the timber cribs, having been towed into its approximate position the previous fall and anchored by cables to stone-filled crates weighing about 15 tons apiece, was caught by a combination of tide, wind, and a fast-moving river swollen by the spring runoff, and it was dragged downstream, stone anchors and all. Neither the weight of the anchors nor the power of two steam tugs could prevent it from being swept almost three miles before a reversal of the tide finally brought it to a stop. For almost a month, O'Rourke's efforts to bring the big crib back upstream met with one disheartening failure after another. Finally, a combination of new piling driven alongside the anchors and a modest respite from the forces of nature allowed the builders to restore the crib to its position and to start their dredging for the next concrete pier. It was an experience that tried O'Rourke's skills and patience, but apparently strengthened his resolve, and by summer things were moving forward again. Before the summer of 1887 was very far along, the Jersey City firm of Ross, Sanford and Baird had begun driving the first of the wood-pile underpinnings for the two truss spans that required falsework. The forest of timber scaffolding would grow as tall as a ten-story building.

During much of the year, the docks at Poughkeepsie had been groaning under the increasing weight of fabricated steel for the trusses and towers, and more of it was coming along every day from Union's shops. In October, barges began towing some of it out into the river, where cranes and ironworkers would begin the work of assembling and erecting it. Raising steel sections as heavy as 20 tons apiece and guiding them into prepared locations, the ironworkers would within only thirty-two working days produce the long-awaited view of the first of the giant bridge trusses on the Poughkeepsie horizon.

Fifteen hundred men were working at the site during the fall of 1887, when that first truss section was completed. Two hundred twelve feet above the river, cranes were crawling out on steel they had just erected to extend the cantilevered arms from both directions and to place connecting trusses between them. East and west of the shore towers there was more of the same. Approach viaducts were extending the bridge's length to 6,768 feet, earning Poughkeepsie (for about a year) the right to boast that its bridge was

FIGURE 6 This photograph taken during construction of the Poughkeepsie Railroad Bridge shows the extensive falsework under the anchor span framing, adjacent to the shipping channel. The cantilvered span allowed the channel to remain clear. Photograph courtesy of Adriance Memorial Library, Poughkeepsie, NY, Local History Collection.

the longest in the world. By the middle of 1888 the superstructure of the bridge was approaching completion, and by the end of the year removal of the temporary trestles, completion of a set of railroad tracks on the bridge deck, and application of couple of coats of paint would bring the seventeen-year effort to a successful end.[23]

Rival plans for nearby bridges faded quickly after Poughkeepsie's opening. Near Peekskill, there was no further effective support for a bridge until well into the next century, when the Harriman family would successfully propose one at Bear Mountain. Storm King's advocates, who had never been able to secure a charter from the state, gave up when their principal sponsor defected to the Poughkeepsie camp.

Construction of the bridge at Poughkeepsie was a colossal and historic achievement, of course, but it came with a last disappointing twist. Commercial constraints and conflicts had in fact delayed and in some cases prevented the bridge company's management from establishing the connections with regional railroads that were needed for economic viability. When the bridge opened, it didn't have a single such connection. The first train to cross had to gain access by temporary track, and it was unable to

FIGURE 7 The completed Poughkeepsie Railroad Bridge. Photograph courtesy of Adriance Memorial Library, Poughkeepsie, NY, Local History Collection.

do anything more than shuttle from one end of the bridge to the other. It would be another six months before the new bridge would connect with railroads coming from and heading toward distant places, and even after that, the bridge would never be a resounding commercial success. The railroads that did rely on the Poughkeepsie bridge would have continuing struggles of their own with the vagaries of regional economics and competition, and most of them would ultimately disappear into the bleakness that overtook the declining American railroad system later in the twentieth century. After a few changes of ownership, the bridge became a property of the New Haven Railroad, which was itself destined for bankruptcy. There were still occasional flashes of success, but little prosperity, and when a stubborn fire damaged the bridge badly in 1974, its owners declined to restore it. It's probably fair to say that although it has an important place in the history of America's great bridges, the Poughkeepsie Railroad Bridge never achieved such a status in the history of its railroads.[24]

Among the railroad bridges that cross the Hudson, the bridge at Poughkeepsie was a later effort, but it wasn't the last. That distinction goes to the Castleton bridge, which crosses about ten miles south of Albany. Completed in 1924, the Castleton was no minor bridge. It boasts a high and dramatic river section, and its approaches required a full mile of structure, but it was built essentially to carry freight trains and for that reason never attracted as much interest or attention as bridges that were being crossed regularly by large numbers of persons living in or traveling through the Hudson Valley.

Proposals for adding a Hudson River crossing at Castleton were starting to be heard as early as 1910,[25] when the growing volume of rail traffic routed through Albany was beginning to be slowed by a bottleneck there. Increasingly, three conditions were slowing things down. One was simply the increase in rail volume passing through the Albany Gateway, more than four thousand railroad cars on a busy day. At least as serious was the effect of the steep grade that westbound trains had to climb in the eight miles just west of West Albany, a stretch that often demanded the use of supplementary engines and sometimes required dividing trains into multiple sections to get them up the hill. And not the least of the reasons for the slowing of rail traffic at Albany was the frequency with which the busy drawbridges across the Hudson had to be opened for river traffic to pass.

To some supporters, a "cut-off" at Castleton that would bypass Albany and sharply reduce the city's rail traffic seemed a natural and reasonable solution to the bottleneck, but not everyone agreed. The plan that evolved brought into existence the New York Central's Hudson River Connecting Railroad, a whole new system that included twenty-eight miles of newly laid double-track railroad, allowing eastbound and westbound trains to bypass the state capital entirely by traveling a diagonal route south and west of Albany and crossing the river on a bridge at Castleton. The centerpiece of what became a $20 million system would be a state-of-the-art classification terminal and service facility at Selkirk, which promised to be the most modern in the world.

But most people working at the old West Albany yards lived close to where they worked and didn't like the new plan a bit, fearing (not without good reason) that many of them would lose their jobs or be forced to move to Selkirk.[26] The skippers of Hudson River vessels didn't think much of the idea either, and they were not about to endorse a plan that involved construction of a bridge across the Hudson at an unusually narrow and somewhat treacherous location. Of course, many people in Albany opposed the plan for a variety of other reasons, sometimes just because they sympathized with the railroad workers or with the ship captains, but more often because they (correctly) anticipated a substantial decline in local tax revenues and loss of local jobs.

But the plan for the Castleton Cutoff had too much going for it, including the vigorous support of Alfred H. Smith, the capable and influential senior executive of the New York Central who would become the company's president in 1914. Over the next few years Smith would overcome or disregard most of the commercial and political opposition to the plan, and early in 1921 major contracts were being awarded for the preparatory grading, drainage, and railroad construction work and for erection of the vast new 8,500-car yard at Selkirk.[27] Later that year a contract for the main grading and drainage work and for all the actual bridge construction except its steel superstructure was awarded to Walsh Construction Company, a well-established Iowa-based railroad builder that would later go on (in a joint venture with others) to build the Grand Coulee Dam and, still later, to achieve national prominence as a builder of urban skyscrapers. The 23,000 tons of structural steel required for the bridge would be fabricated in Pittsburgh and erected by Bethlehem Steel's McClintic Marshall subsidiary under a direct contract with the railroad.

Actual clearing and grading work got underway early in 1922. Between then and the end of 1923, concrete pedestals for the 1,250-foot-long westerly viaduct, for the 3,000-foot-long easterly viaduct, and for the three massive river foundations and their stone-faced piers were completed and made ready for steel. The three concrete river piers, which averaged close to 160 feet in height, were founded on bedrock and built within steel cofferdams, but two of them needed the additional benefit of reinforced concrete and pneumatic caissons. Almost thirty-five years had elapsed since the more primitive timber cofferdam approach had been used at Poughkeepsie. Around the middle of 1923, the first structural steel began arriving at the site, and by the end of the year the steel towers for the viaducts and the steel framing for their decks were approaching completion, leaving the installation of the concrete deck and the track work to be done in the spring.

The work of assembling and erecting the huge trusses that would take the double-track system across the river was started early in 1924, using steel (instead of wood) for the falsework and sequencing the installation to allow for one of the two river channels to be left open to marine traffic while the other was obstructed. The 408-foot-long easterly trusses were erected

A. H. Smith Memorial Bridge Over the Hudson River

FIGURE 8 The Alfred H. Smith Memorial Bridge, most frequently called the Castleton Cutoff. Photograph courtesy of the Richard J. Barrett Collection, Albany, NY.

first, and as soon as the falsework under them could be safely removed and installed in the channel under the longer span, erection of the 600-foot westerly trusses was started. The longer trusses were about 100 feet deep, and the shorter trusses were a few feet shallower.

By November 1924 concrete work for the whole deck was in place, and the tracks had been carried across the river. On the twentieth of that month, New York Governor Alfred E. Smith headed a slate of dignitaries in a ceremony that opened the new bridge, formally naming it the Alfred H. Smith Memorial Bridge, after the man who had been its leading advocate for ten years. That Smith, whose name has often been confused with the almost identical name of the governor, had been killed in a horseback-riding accident in New York's Central Park only a few months before the bridge was finished.

The Castleton Bridge and the system of which it was a part performed well, accomplishing just about everything their advocates had expected. But they are still not popular in Albany, where some old-timers remain resentful that the project effectively ended the useful life of the New York Central's West Albany yards, eliminating hundreds of jobs there and in the surrounding area and damaging the economic well-being of Albany itself.[28]

CHAPTER 5

The Railroad Tunnels

BY THE BEGINNING OF THE TWENTIETH CENTURY, bridge building on the Hudson River had shut down for a while. The crossings that had started with a modest wooden span at Waterford in 1804 had by the early 1890s become a series of bridges that laced together the banks of the upper and middle sections of the river, expanding commerce and increasing prosperity in much of the Hudson Valley. But since completion of the bridge at Poughkeepsie in 1888 there had been signs that the engine that had been driving almost a century of bridge building might have run out of steam.

The motivation behind almost all the bridges that had been built was the need to carry railroads across the river. Once that need had been met at Poughkeepsie, investors just didn't see any promising places for new bridges anywhere between there and New York City. Not that there hadn't been plenty of growth and development along that approximately seventy-five miles of river between Poughkeepsie and New York, but the need to get across anywhere along that stretch was essentially a local one that was being effectively met by well-established ferry systems.[1] Between the new bridge at Poughkeepsie and the northern boundary of New York City, there wasn't a single place where railroads had reached the water's edge (or were planning to reach it) and were looking for a way to get across.

The picture was radically different along the stretch of river that separates New Jersey from New York. New York City had grown explosively. By the late nineteenth century, having absorbed the neighboring city of Brooklyn, it was home to a population of about 3.4 million, up from about 1 million only twenty-five years earlier and more than ten times the combined populations of all the upriver towns where bridges had already been built.[2] By the turn of the twentieth century, it had become the largest and most important city on the continent, a center of domestic and international trade, finance, manufacturing, science, and publishing, awash in theaters and new museums and rich in talent. By then, New York was developing and manufacturing much of what the rest of the country would need to move into the new century, and it was doing some of it in buildings that had elevators and telephones. Everything was very much up to date in New York, but no one had yet built a bridge across the Hudson to New Jersey.

That certainly wasn't because there were no railroads positioned to cross the river at New York. The Hudson River coast of New Jersey was almost always crowded with long trains that had brought passengers and freight from all over the country but were prevented from continuing on into or beyond New York by the impossibility of crossing the river. There were ten of those big railroads, including such giants as the Pennsylvania, the Lackawanna, and the Erie, but for the most part their managements accepted with apparent contentment an arrangement that allowed their huge, eastbound trains to travel only as far as the ornate rail and ferry terminals that lined the Jersey shore of the Hudson. There, passengers and freight transferred to ferryboats that took everyone and every thing across the river. That same fleet of ferryboats then brought back westbound passengers and freight for transfer to westbound trains in New Jersey. It was an archaic system that was years behind the times; inherently slow and inconvenient, it was characterized by often rough and sometimes dangerous trips across the river and by a low degree of reliability that is not surprising where ice, fog, wind, and other natural enemies of modern marine transportation can wreak havoc with schedules. But for the most part it was a system that was accepted by the traveling public without much protest. The elegant terminals and most of the ferries were owned or controlled by the railroads themselves. Given the high level of popular acceptance and the profitability of the system, there was little incentive to find a better way.

That complacency didn't discourage everyone, however, and ideas for alternative approaches had been surfacing fairly regularly over the course of the nineteenth century, ever since John Stevens had proposed his much-maligned pontoon bridge back in 1805. But the problem was far from simple. The river was over a mile wide at New York, so the cost of a bridge there was bound to be high, the tidal range was large, the currents were powerful, and rock for supporting bridge foundations was a long way down. The harbor was already the busiest on the continent, and the shipping companies were as determined to resist interference with river traffic as they had been when the upstate bridges were first proposed.

Bridges, of course, were not the only alternative to ferries. As the prospect of a bridge gave signs of waning, a certain amount of talk about building a tunnel began to be heard. Tunnels have a history that goes back to the Romans and earlier. For a thousand years after the Romans lost power nothing much was built, but eventually the Moors picked up the thread with their tunnels in Spain. In the seventeenth century, the French built the Canal du Midi, an important canal that included a 500-foot-long tunnel, reviving interest in tunnel building for a while. About 150 years later British tunnel builders led by the Frenchman Marc Isambard Brunel managed to drive a primitive tunneling shield 1,200 feet through treacherous soils under the Thames River to complete the world's first major subaqueous river crossing in 1843. That spectacular first use of a tunnel shield initiated the modern approach to building tunnels under major rivers, notwithstanding the eighteen years it took to complete the Thames Tunnel. Within another 50 years the British would build five more tunnels under the Thames. By the late nineteenth century the technology was beginning to find its way across the Atlantic to America, where canals and then railroads (neither of which could tolerate steep grades) created a real need for tunnels.

Many of the early tunnels in America were done as "cut-and-cover" construction, especially along the almost-level canal routes and some of the railroad routes. They were often shallow and could be built in trenches excavated from the surface of adjacent ground and then backfilled, so their construction was not—strictly speaking—"tunneling." More than a few of the other efforts, like the famous Hoosac Tunnel in Massachusetts, which required about twenty-three years to complete and took 195 lives, were tediously chiseled through hard rock, using methods similar to those employed in mining. Those early American tunnels certainly met the needs of the period, but building them was for the most part slow and primitive work, and they didn't provide much of a model to guide later builders who wanted to drive tunnels through the alluvial soils under the Hudson River. The sometimes nasty business of driving tunnels under rivers would still be a dangerous and complex task well into the nineteenth century.

One man who wasn't intimidated by the dangers and challenges of tunneling under the Hudson was a self-styled "practical engineer" and sometime adventurer/entrepreneur named DeWitt Clinton Haskin. Born in New

York State in 1824, about when construction of the earlier DeWitt Clinton's great Erie Canal was being completed, Haskin had in 1849 joined thousands of others in a frenzied rush to exploit the discovery of gold in California. As things turned out, he was one of the lucky ones—not because he plucked gold from a stream or dug it from a mine, but because he was able to convert a northern California inn to a successful hotel and then to operate it profitably for the rest of the boom years. He took his money and went on to other successful commercial ventures out West, including some mining in Utah and the construction of a short railroad near San Francisco that he later profitably sold to the Central Pacific. By 1872 he was back in New York, a good deal richer than when he'd left and still energetic, ambitious, and imaginative.[3]

It was on that return trip to New York that Haskin first thought about a tunnel under the Hudson. The ferryboat taking him across the river from New Jersey had become fogbound along the way, and during the hours that elapsed before the boat was secured on the New York side, Haskin had plenty of time to think about how successful such a tunnel might be. By the time he disembarked, he'd laid a plan for exploring the idea further.

Tunneling, mining, and even railroad building are not altogether different activities, and Haskin already knew something about mining and railroad building. On the other hand, he realized that burrowing through the silty morass under the Hudson's waters was not entirely like what he'd been doing, and it properly worried him. He explored the subject in some detail, probing available tunneling and other marine construction technologies and paying special attention to an idea suggested by a technique that was being used for some of the new deep-water bridge foundations. On some of those projects, compressed air was used to keep the excavations dry, pressurizing the working spaces to hold back the water until the substructure work could be finished. After visiting some of those projects himself during 1872 and 1873, Haskin had seen enough. Convinced that maintaining an elevated level of air pressure within the forward chamber of a tunnel could keep it dry until a permanent, watertight liner could be put in place, and that a controlled airlock for entering and leaving the pressurized space would protect the workmen, he sprang into action. By the middle of 1874 he had secured the necessary charter, permits, and $10 million in subscriptions for shares and bonds in his new Hudson Tunnel Railroad Company, and by fall he had begun work on a vertical shaft just inside the New Jersey bulkhead line. It was located to provide construction access to the western end of a tunnel he planned for connecting Fifteenth Street in Hoboken, New Jersey, with a landing site at Morton Street in New York.

Before the 60-foot-deep access shaft could be completed, the project was abruptly halted by an injunction obtained by the Delaware, Lackawanna and West Railroad, under whose land Haskin was planning to drive his tunnel. It was a serious and (surprisingly) unanticipated defeat for Haskin, and it would be five years before his arguments against the injunction would prevail. By then the cost of fighting it in court and the impact of the lost time had weakened him financially and physically regaining the original momentum would prove difficult. Haskin was able to reassemble much of his original staff and crew, but he turned to the engineering firm of Spielman and Bush to manage the work. It rejected his original plan for a double-track system in a single 26-foot-wide-by-24-foot-high tube in favor of building two single-track, 18-foot-wide-by-16-foot-high elliptical tubes. By the following summer, the company had put in place several hundred feet of well-constructed, brick-lined tunnel.[4]

Most of what Spielman and Bush had done was in the north tunnel, which is where disaster struck on 21 July 1880. Air pressure in the working section blew a hole in the muddy silt of the river bottom, just above the place where men were working, allowing tons of material and the river itself to pour into the tunnel. During the precious minutes that followed discovery of the breach, a few desperate men managed to pry open the watertight door that separated the working area from the airlock, and they were able to escape through it to safety. But in the ensuing chaos a heavy section of temporary shoring collapsed against the escape door, crushing a man to death, rendering the door impassable, and condemning to a watery grave the nineteen other men who were still out in the working area of the tunnel.[5]

It took approximately six months to repair the blowout and seal it. Afterward, Haskin made a few more attempts to continue the work, but he was running out of money and vigor. Within a couple of years his company was bankrupt, and he retired to live another eighteen years before dying impoverished and blind in 1900. In the meantime, a committee of the bondholders managed to stir things up again in 1888. They reorganized, raised some additional working capital, mostly from British sources, and brought from England the venerable construction firm of S. Pearson and Sons to continue the work.

The Pearsons, who had been in business since 1844, had built some of the most important structures in the British Isles, and they went right to work on continuing the Haskin tunnel. They brought new enthusiasm and talent to the project, together with some important technology. They elected to line the tunnel with a system of heavy cast-iron plates, segmented and shaped to fit the tunnel's interior and bolted together through flanged joints, eliminating the need for the masonry liner, saving money and speeding the work. More important, the Pearsons brought with them the newly developed Greathead Shield, a gigantic tunneling device that looked a little like the front end of a locomotive and had been named for one of the first engineers to use it effectively. Its face was its cutting edge; once positioned at the head of a tunnel under construction, it could be forced ahead by a system of powerful hydraulic jacks, in increments of about 30 inches, through the soft soil of the river bottom. Small openings in the face of the shield were designed to allow the muddy soil to ooze back into the working space of the tunnel, where laborers could shovel it into wheel-mounted containers that were guided on tracks back to the access shaft for emptying above ground.

Pearson made good progress, despite a few blowouts, and was able to advance the work to within less than 2,000 feet of its destination, but the owners of the tunnel company ran out of money again in 1892 and were unable to raise enough to finish the job. The hard luck that seemed to have gripped the project at its very beginning just wouldn't let go. By the end of the year the tunnel company was in bankruptcy, and the Pearsons were off the job.[6]

At about that same time, a thirty-year-old Georgia-born lawyer named William Gibbs McAdoo arrived in New York with his young family, seeking a career that would be broader and more enriching than the one he had started only a few years earlier in Chattanooga, Tennessee. Intelligent and charming in the manner of a cultivated southern gentleman, he was attracted by the scale and pace of life in a major American city like New York. He would distinguish himself there, establishing a reputation for integrity and competence. After a couple of busy decades he would move on to Washington to become secretary of the Treasury in Woodrow Wilson's cabinet. During the years following that prestigious appointment, McAdoo would at various times serve as a founder of the Federal Reserve System, chairman of the Federal Reserve Board, and director general of railroads. Still later in his long life, he'd be elected as a U.S. senator from California.

Back in those early years in New York, when everything there was new to him, McAdoo divided his energies between establishing a law practice and exploring what was to him the uniquely fascinating commerce of the big city. Among many other things, he wondered why there was no way to cross the river between New Jersey and New York except by ferryboat, and he thought more than a little about what it would take to build a tunnel and run an electrically driven railroad through it. He had never heard about Haskin's attempt, and his own ideas about building a tunnel would probably never have come to anything at all if it hadn't been for a chance professional association with a lawyer named John R. Dos Passos, who still owned some of the nearly worthless bonds of the old tunnel company.[7]

Early in 1902 Dos Passos introduced McAdoo to a capable engineer named Charles M. Jacobs, a distinguished fifty-two-year-old Englishman who had a dozen years earlier come to New York to provide advice to the young Long Island Railroad (which in 1900 had been acquired by the Pennsylvania Railroad) about building a tunnel under the East River. Within a short time Jacobs's advice was being sought by almost everyone in New York who had a marine structure to build or improve, and McAdoo was brought to the head of the waiting line by Dos Passos.

Educated at Cambridge University and in the shipbuilding yards of his native Hull in Yorkshire, Jacobs had by the late nineteenth century become well respected in England for his expertise in the complex business of building tunnels and other marine works, but it would be later, in the United States, that he would become acknowledged as a uniquely eminent expert in the field. By the time he and McAdoo met in 1902 to talk about the Hudson River project, Jacobs had designed and built the first tunnel to penetrate the treacherous and often unpredictable soils that underlie the East River, a structure that would carry a 36-inch gas main and an adjacent narrow-gauge servicing track between New York City and Long Island. He and his recently acquired partner, J. Vipond Davies, made no effort to conceal their interest in McAdoo's idea of building a tunnel under the Hudson that would be larger in diameter, twice as long, in some places twice as deep, and for the most part at least twice as difficult as what they had built under the East River.[8]

When it came to deciding whether or not the formidable and costly task of trying to complete work on the twice-abandoned Haskin project made a bit of economic sense, McAdoo soon realized that there was a limit to how much he could learn from simply talking about it with Jacobs, and within a few days of their meeting the two of them had descended the access shaft into the abandoned tunnel to see just what was what. Dressed in rubber hip boots and yellow slickers, and carrying oil lanterns to penetrate the blackness of the place, they trudged almost two-thirds of a mile along the muddy bottom of the unfinished tunnel before reaching the Greathead Shield, which lay in their path like a mortally wounded giant. Jacobs probed the big machine deliberately for a long time, like a careful doctor examining an injured patient, and finally pronounced the shield salvageable with a modest investment of work. In his view, construction of the tunnel should be resumed right away.

That was good enough for McAdoo, and it didn't take long for his interest in the project to become his passion. He was not entirely without experience in the business of electrically operated light railways. In the early years of his legal practice he engaged in a sustained but ultimately unsuccessful (and costly) effort to convert the mule-drawn street railways of Knoxville, Tennessee, to electric power. Now, he saw a light rail system under the Hudson as the ideal candidate for electrification for a railway. His grand strategy was to develop such an electrically operated, interurban railway to transport travelers and commuters between New Jersey and New York. Almost at once, he began to focus most of his energy on that idea, first establishing the New York and New Jersey Railroad Company, then issuing $5 million worth of bonds and personally selling them to some of the most powerful people in the Wall Street community. Next, he proceeded to acquire the rights of the old tunnel company from its bondholders for the bargain price of $350,000. By the middle of 1902 he had wound up what remained of his law practice, and within another few months the work of building the tunnel had been resumed under the personal direction of Charles Jacobs.[9]

Meanwhile, McAdoo and Jacobs hadn't been the only ones thinking about ways of getting people across the Hudson between New York and New Jersey, but few of them were thinking about tunnels. Most were men who designed and built bridges. No fewer than seven well-developed bridge schemes had been publicized in New York between about 1885 and about 1900, including a cantilever design from the Union Bridge Company, builders of the Poughkeepsie Railroad Bridge, and a proposal for a suspension bridge from a board of engineers appointed by the War Department. But none of the late nineteenth-century ideas for bridges would be as effective in keeping alive the idea of a bridge between New York and New Jersey as the plan for a suspension bridge "somewhere between 14th Street and 28th Street" that had been presented to a meeting of the American

Society of Civil Engineers (ASCE) early in 1888 by an engineer named Gustav Lindenthal.

The Austrian-born, thirty-eight-year-old Lindenthal had supplemented a modest early education with a few years of solid bridge-building experience in the engineering departments of some of central Europe's railroads before emigrating to the United States around 1875. After working in the Philadelphia area for a few years, he had moved on to Pittsburgh, where he established an engineering practice that would be strikingly productive during the relatively few years he was there. In addition to providing engineering services to some of the railroads and towns of the area, Lindenthal's office had during that brief period been able to design and build Pittsburgh's Seventh Street suspension bridge across the Allegheny and its innovative Smithfield Street truss bridge across the Monongahela, a replacement for an earlier Roebling bridge. These were regarded as two of the most important structures built in America during the 1880s, but by about 1885 Lindenthal had apparently tired of the relatively modest challenges offered by Pittsburgh's narrow rivers and decided to prove himself on New York's wider and more challenging Hudson.[10]

Most of the engineers who crowded into that historic meeting of ASCE's New York chapter on a winter evening in 1888 probably knew little about Lindenthal and had learned only recently that he was in town, but they understood that he was a fellow who was likely to have something interesting to say. They were not disappointed. What he showed and discussed with them (for three and a half hours) was his design for a Hudson River suspension bridge of three spans, the longest of them about 2,850 feet, almost twice the span of the recently completed Brooklyn Bridge. It would carry six sets of railroad tracks and terminate in a glorious new structure near Twenty-third Street in Manhattan. The whole project, Lindenthal told a reporter after the meeting, might cost as much as $50 million, including the cost of the land and the terminal structure.[11]

As it turned out, Lindenthal, for a year or more, had been having private discussions about his bridge ideas with Samuel Rea, then a mid-level executive (later president) of the Pennsylvania Railroad, which had for years been chafing at its inability to bring its fleet of trains into the city of New York.[12] The Pennsylvania had the potential for becoming Lindenthal's principal patron and customer, and Rea had given him serious reason to expect that the company would support his plan. But just before the 1888 meeting, Rea suddenly quit his job at the Pennsylvania to take what he saw as a better opportunity with the Baltimore and Ohio, leaving Lindenthal without a reliable sponsor and lending some urgency to his presentation to the ASCE.

This disappointing setback turned out to be a relatively brief one. By 1892 Rea had soured on the Baltimore and Ohio and returned to the Pennsylvania, where he had started years earlier as a rodman in a survey party. His enthusiasm and support for Lindenthal's bridge remained undiminished. The Pennsy's board would soon formalize its support for the Lindenthal project with a resolution to guarantee $800,000 in annual fees for using the bridge, which by then had grown to include two levels, fourteen sets of train and trolley tracks, numerous carriage and pedestrian lanes, and a grandiose terminal on the West Side of New York. The estimated cost had increased to $100 million.

But there was a fatal catch in the Pennsylvania's guarantee. Even the mighty Pennsy did not want to carry the whole cost of such a project alone, so the board's promise was contingent on Lindenthal's ability to secure similar commitments from the other railroads that would use the bridge. It seemed an entirely reasonable condition, but it proved to be an impossible one. The other railroads, aware that the Pennsylvania would be compelled to give them access to a federally chartered bridge, were simply unwilling to make any guarantees. Even pressure from the Pennsylvania's top executives didn't help.[13]

The resolve of the Pennsylvania's board weakened, and by the time construction on McAdoo's tunnel got underway late in 1902, speculation about whether Lindenthal's bridge would ever be built had become moot. Not only had the Pennsylvania decided against the bridge, it had decided to build tunnels instead. Alexander Cassatt, recalled to the company as president after a seventeen-year retirement and recently returned from Europe, had experienced an epiphany. The French had brought the technology of electric traction to a new level, and Cassatt had witnessed the operation of electrically powered trains that were every bit the length and weight of the Pennsylvania's. He decided (and convinced his board) to abandon the idea of a bridge and to build tunnels instead.[14] The new electric locomotives, he convinced the board members, were able to match or exceed the power of steam locomotives, and they didn't produce the smoke that for years had made the bridge immensely more attractive than a tunnel.[15]

Just when McAdoo first learned of the Pennsylvania's decision, or whether he ever thought of the railroad's tunnels as a threat to his own project, is not really clear. There is not a word in his 1931 memoirs to suggest that the railroad's actions ever worried him at all. The Pennsylvania's tunnels would be substantially larger than his, built to allow its huge trains and those of competitors to pass directly into and out of a new terminal in New York. McAdoo's interurban railway would handle passengers who were for the most part just traveling between the towns of eastern New Jersey and downtown New York. The evidence is strong that McAdoo never thought of his own system as anything more than a light railway whose time had come. In later years, in fact, the commuting population would enable the interurban system to survive while the customer base of the big railroads declined.

Construction of "McAdoo's tunnel," as it was increasingly being called in the press (over his own self-effacing objections), resumed in the fall of 1902 in the north tunnel, just where Pearson had abandoned it. Jacobs, seeing no need to engage a general contractor to replace Pearson, had simply taken charge of the work himself, hiring staff and purchasing materials and equipment directly. Progress was slow, so he built a second access shaft at the New York shore to allow for boring simultaneously from both ends of the tunnel. Using the original shield as a model, he arranged for a second one to be fabricated in New York, and by the beginning of 1903 two crews were approaching each other under the river.

Of course, such work can hardly be expected to be uneventful. No remedy was ever found for a troublesome, leaking joint in the tunnel lining, just behind the eastbound shield, where the skirt of the shield met the permanent liner of the tunnel, and it was difficult to prevent some silt from penetrating the tunnel. Whenever the forward movement of the shield slowed, the problem was exacerbated. As soon as a stall occurred, the force of the high air pressure being maintained in the tunnel heading began to widen the opening, admitting more silt and sometimes even admitting water from the river above. Keeping crews at work around the clock, seven days a week, tended to avoid most of the slowing or stopping of the shield's forward movement, but it wasn't a perfect solution. On one occasion a stall lasted long enough to allow the pressure to make the opening dangerously wide, flooding the tunnel. Everyone managed to escape, but plugging the leak had to be done by dumping hundreds of tons of silt from barges positioned in the Hudson above the tunnel, a task that demanded some elegant mathematics and nimble surveying, and one that generated a costly and unwelcome delay. Before all the water had been pumped out of the tunnel, it was realized that fish had entered through the leaking joint, and at least one fish was taken by an adventurous workman, probably the only one ever caught in the Hudson from below its bed. Soon afterward, an intimidating ledge of rock that lay in the otherwise soggy path of the shield proved to be another frustrating source of trouble. Seven hundred feet long, it had to be tediously drilled and blasted, delaying the progress of the work by almost eleven months.

Things weren't all bad, and some of the lost time was recovered. Jacobs had made an important improvement in his tunneling technique, tripling the speed at which the shield could be advanced. By increasing the number and individual capacities of the hydraulic rams he was using to drive the shield, he found he could advance it more rapidly through the silt and that he could do it with the access doors fully closed. The added power of the rams made it possible for the shield to force the silt out of its way as it advanced, keeping all the material out of the tunnel heading and eliminating the labor-intensive and time-consuming work of loading it into carts, transporting it to the access shaft, elevating it to the surface, and carting it away.[16]

While all this was going on, McAdoo was working out the details of his broader concept. From the beginning, he had believed that a second set of tubes would eventually be necessary, farther downtown, and by early 1903 he had decided that such a pair of "downtown" tubes should be built about a mile south of the uptown tubes. His idea was to position them to connect the Wall Street area of New York with the southerly end of the Jersey City waterfront, and to connect the downtown and uptown tunnels with a short, north-south subway spur on the Jersey side. In New Jersey, the busy terminals of the Pennsylvania, the Erie, and the Lackawanna Railroads were all lined up along that stretch of the waterfront, and McAdoo's subway would be positioned perfectly to pick up any passenger who arrived on one of those big trains and wanted to cross into the Wall Street area. When work in the downtown tunnel was started in the spring of 1903 (using the heavy-duty rams and newly fabricated shields), the goal was to catch up to the north tunnels, but that proved to be too difficult a challenge. The two north tunnel shields came face to face in March 1904 (with what McAdoo called a watchmaker's preciseness) and the two shields in the south tunnel followed in September 1905.

McAdoo's expansive ideas didn't end with a second tunnel. They went on to include plans for connecting his interurban railway with New York's fledgling subway system, and perhaps even extending it westward into New

Jersey, toward Newark. Even more daring, in 1903 McAdoo started to think seriously about building a multilevel underground terminal at the New York end of the tunnels, as well as an approximately one-million-square-foot income-producing office complex above it in the form of two soaring twenty-two-story towers. The complex would become the largest of its time.

Implementing such ideas had to start with some kind of agreement with the Pennsylvania Railroad, which by then had announced and begun work on its plan to enter New York through a system of its own tunnels, north of McAdoo's. The Pennsylvania's railroad and ferry terminal was the most southerly of the three big terminals that lay along the Jersey waterfront, and before McAdoo could get construction of his downtown project underway, he was going to have to get permission from the Pennsylvania to drive his tunnel under its property. The terminal had been a busy and profitable facility for the railroad for many years, handling thousands of passengers every day. McAdoo's interurban railway was certain to take away some of its business.

But in McAdoo's thinking, his request would be viewed favorably by anyone thinking in terms of the Pennsylvania's long-term objectives. It wouldn't be long, he reasoned, before the Pennsylvania and the other railroads would want to abandon the awkwardness of maintaining a Jersey City station in favor of new routes that would bring their big, high-speed trains directly from Newark to and under the Hudson through the Pennsylvania's new tunnels and on into New York's Pennsylvania Station. That eventuality would leave in the lurch a fair number of passengers whose destination was the Wall Street area of lower New York, nowhere near the new uptown terminal, and it was McAdoo's guess that the Pennsylvania's Alexander Cassatt would have little taste for ignoring such a loyal and deserving population. He was right. It took only about an hour's discussion for Cassatt to agree that McAdoo could drive his tunnel under the Pennsylvania's Jersey property and that his interurban railway should be linked to the Pennsylvania's somewhere along the route to Newark, where it would pick up those Pennsylvania passengers who were headed for Wall Street. The trip through Jersey City, across the Hudson through the McAdoo tunnels, and on into the Wall Street district of New York would be made without anyone's having to worry about whether or not the ferries were running.

Building the downtown tunnels started a few months after McAdoo's meeting with Cassatt. For the most part it was a repeat of the uptown project, but a bit easier and faster, lessons from the first project having been applied to the second. McAdoo established the Hudson and Manhattan Railway Company to build the downtown tunnels, and a few years later that company and the original New York and Jersey Railroad combined to form the Hudson and Manhattan Railroad, a company that would years later be absorbed into the Port Authority Trans-Hudson system, ultimately called PATH. The opening of the uptown tunnels, elaborately celebrated by such dignitaries as the governors of New York and New Jersey and joined by telegraph to President Theodore Roosevelt in Washington, occurred on the twenty-fifth of February 1908, and was followed about a year later by the opening of the downtown tunnels.[17]

That whole first decade of the twentieth century was an intense one, with immigration surging, new ideas and new technologies abounding, and a whole new culture emerging. It's hard to imagine any person of that period who could have been busier than Charles Jacobs. Only a few months before starting his big job for McAdoo, Jacobs had accepted Alexander Cassatt's invitation to assume a similar and largely simultaneous role in the Pennsylvania's hugely ambitious program to build its own tunnels under the Hudson.

What the Pennsylvania was planning went well beyond tunnels. In Cassatt's concept, the railroad's main line would still pass through Newark, but instead of continuing to Jersey City, it would veer northward (as McAdoo had expected) along a new embankment to be placed above the Hackensack meadows; then it would continue on through a tunnel that would be drilled and blasted through the solid granite of Bergen Hill. When the line reached the Hudson, it would pass under it through one of two single-track tunnels, emerging in an elegant new terminal in Manhattan. From there, in the Cassatt plan, the Pennsylvania's trains would continue east under Manhattan itself, cross under the East River through another set of tunnels, and emerge at Sunnyside, on Long Island, where the railroad would build a new servicing facility capable of handling 1,500 railroad cars at one time.

And all that was only the first phase of what Cassatt saw as the destiny of the Pennsylvania Railroad. He and his colleagues had for years wanted their trains to be able to travel all the way up the coast into the New England states. For that final stage of his plan, he anticipated a bridge at Hell Gate to carry the railroad's passengers and freight from Long Island back to the mainland north of Manhattan, positioned to continue on up into New England. That phase of the plan would be deferred for a dozen years or so, but engineering and design work for the rest was to be started immediately, in 1902.

To manage all this work, Cassatt established a board of engineers that would "pass on the practicability of the undertaking, determine . . . the best plan for carrying it out, make a careful estimate of its cost . . . and exercise general supervision of its construction." He appointed Jacobs to that board early in 1902, along with four other eminent engineers: as chairman, Colonel (later General) Charles W. Raymond of the U.S. Army Corps of Engineers; Gustav Lindenthal, the disappointed bridge designer who would resign in late 1903 to become commissioner of bridges in New York City; William H. Brown, who had served as chief engineer of the Pennsylvania Railroad for twenty-eight years; and Alfred Noble, the former president of the American Society of Civil Engineers. The addition of George Gibbs only a few months later gave the board an especially valuable member. Educated in electrical engineering at Stevens College, Gibbs was a gifted practitioner of all the engineering disciplines and had cut his teeth in the employ of George Westinghouse. He would manage the formidable tasks of electrifying the tunnels and their approaches, developing the electrical system for the new terminal building, and even designing the new steel cars that the Pennsylvania would use.

Lindenthal, whose considerable and widely acknowledged expertise as a bridge designer was not likely to contribute much to what this board had been asked to do, might have been appointed because Cassatt felt a measure of obligation to him for his years of work on the abandoned bridge option. Lindenthal would renew his relationship with the Pennsylvania a few years later, as an engineer in private practice, when his office would be selected by the railroad to design the Hell Gate Bridge.

Cassatt started things off by telling the members of the engineering board that they should not allow cost to be a governing factor in designing and building the system. Rather, he emphasized that "the main considerations [must be] safety, durability and proper accommodation of the traffic." The engineers apparently heeded his advice scrupulously. By the time it was all finished, the work (not including the Hell Gate phase) would cost about $160 million, expressed in 1909 currency.

The project was divided into four parts: the Meadows Division; the North River Division (the Hudson River between New York and New Jersey was still called the North River in 1902); the Terminal Station Division; and the East River Division. Members of the board (together with the distinguished architectural firm of McKim, Mead and White) would be given responsibility for managing three of those parts, but the job of managing the North

River Division, the one that would include the critical pair of Hudson River tunnels between Weehawken, New Jersey, and the east side of Ninth Avenue in New York, was given to Jacobs.[18]

There was a prodigious amount of engineering to be done before any tunnel construction could be started. Some preliminary work, like establishing base lines and locating tunnel centerlines, had been started as early as 1901, before the engineering board had been formed, but the essential tasks of identifying and exploring the myriad complexities of the project, designing the tunnels and their approaches, and then preparing detailed documents from which contractors would be able to prepare competitive bids still lay ahead when Jacobs assumed his duties in 1902.

Almost immediately, he established a task force to design the access shafts needed at both ends of the tunnels and then awarded early contracts for building them, ensuring that their construction would be well advanced when design progress made it possible to award a contract to build the tunnels themselves in the spring of 1904. Between the demands of the Pennsylvania Railroad project and the concurrent duties of running the McAdoo project, these were busy years for Charles Jacobs.[19]

The challenges of designing the new tunnels were eased a little, of course, by the experience that Jacobs had been accumulating on the McAdoo tunnels, where construction was already underway. But there was one difference between those projects that troubled him. He worried that the Hudson River silt on which the Pennsylvania's tunnels were going to rely for support might become unstable under the greater weight of their larger, more substantially constructed tubes and under stresses that would be imposed by their faster and heavier trains. Right from the beginning, a good deal of attention was focused on this subject, and the possibility of supporting the tunnels on piling driven to solid bearing was explored in detail. Several 30-inch-diameter screw piles were tested and found capable of sustaining loads of 600,000 pounds for a month without showing any subsidence, but it was decided to defer a decision about whether or not to use piling, and openings for such piles were added along the entire length of tunnel floor, at 15-foot intervals, just in case. It was a way of deferring a difficult decision until more information became available, and the construction contract was written accordingly.[20] The tunnel, one newspaper writer observed, might eventually work like an underground bridge, spanning from pile to pile.

There has always been a fairly tight community of engineers and construction workers who follow and work on big river jobs, and it's not unlikely that some of the men working on the Pennsylvania tunnel project late in 1903 included a few veterans of the great Poughkeepsie Railroad Bridge, which had been completed about fifteen years earlier seventy-five miles upstream. If so, some of those veterans would probably have recognized the contractor who in December 1903 submitted the winning bid for the new tunnels. He was John F. O'Rourke, the charming, Irish-born, Cooper Union–educated civil engineer who had headed up construction in Poughkeepsie and endeared himself to workers and to much of the community. O'Rourke had married the daughter of one of the town's leading families and headed off with her to New York, where he established the John F. O'Rourke Construction Company. By 1903, he had done some of the city's most challenging heavy construction work and was well positioned for the big tunnel job. His successful bid for the tunnel contract was good news for Jacobs, who knew the company and its work well.[21]

The contract was signed in May 1904. Although it seemed to many who had been following the process that the main job might finally begin in earnest, there was still plenty of work to be done before sandhogs could begin digging their way toward one another from opposite sides of the river. The tunneling shields had already been redesigned, but their components had still to be fabricated. The shields themselves had to be assembled at the site, and access tunnels had to be excavated through and under the cluttered waterfronts of New York and Weehawken before those 194-ton behemoths could be maneuvered into position to do their jobs. It would be the middle of 1905 before all that difficult and sometimes tedious work was finished.[22]

Even then, there would be problems. Only a few days into the excavation of the river tunnel, the engineers realized that the elevation of the face of the shield had somehow become too high and was rising, and by the end of the work shift, the giant shield was found to be several feet above where it was supposed to be. It was no minor problem, and the engineers addressed it quickly and vigorously. It turned out that the huge mass of silt that needed to be pushed aside was resisting the forward movement of the Pennsylvania's larger shield, pushing it up instead of allowing it to move forward. Opening the shield's doors would solve the problem, but that was a solution no one wanted: it would mean admitting a massive volume of silt into the tunnel, shoveling it into carts, moving it back to the shaft, elevating it, and disposing of it at ground level. A few crude experiments produced a compromise. Allowing an average of about one-third of the excavated material to ooze through a partially opened door or two would reduce the resistance by enough to allow the shield to remain on course. During the rest of the work, opening and closing the shield doors in calculated increments became a new and useful way of controlling the tunnel's level, but it still meant the loss of considerable time and money.[23]

By the middle of 1906, the builders of the Pennsylvania tunnels had experienced and survived most of the delaying and sometimes cruel events that had plagued the builders of the McAdoo tunnels and their failed predecessors: equipment breakdowns, leaks, "blowouts" (including one that required 25,000 bags of cement and a like amount of sand to patch), and sporadic illnesses thought to have been caused by high air pressure in the tunnel. A curious problem that nagged at the engineers, even after they had opened the doors of the shields, was the relatively small, but continuing, periodic rise and fall of the tunnel floor itself, independently of the shield's progress. After a good deal of investigation, the movement was correlated with and attributed to the rise and fall of the tide, but the exact mechanism by which tidal movements affected the tube levels was never really determined. The magnitude of those variations diminished over time, and later it was decided to ignore them as insignificant.[24]

Not every challenge went against the builders. Sometime after the north and south tunnels were holed through in September and November of 1906, respectively, Jacobs and his staff readdressed the still unanswered question of whether or not the screw piles would be required. Working through bore holes that had been made for subsurface investigations under the tunnels, rods were driven down into the bedrock below and used as benchmarks for observing and recording elevations in the tunnel before and after its very heavy, two-foot-thick concrete lining had been installed. The concrete liner, which had not been used in the McAdoo tunnels, was designed primarily to protect the 30-inch-wide cast-iron rings that had been bolted together through their flanges to form the tube itself, and to stiffen the tunnel's resistance to the external pressure of the soil. In fact, a suggestion by the railroad's president had determined its geometry. The concrete liner was shaped to include an integral, continuous, 44-inch-wide concrete bench just one foot below the horizontal axis of the tunnel, on each side, to provide an escape sidewalk for railroad passengers who might in some future emergency need to exit the railroad cars through their windows. The huge volume of concrete generated by that profile took more than a year to form and install, and almost tripled the weight of the tunnels. With that massive volume of concrete in place and no evidence of settlement, the engineers began to feel new confidence that the piling probably wouldn't be needed after all. Precise observations made during and after the installation of the concrete continued to confirm the previously accepted tidal variations, but nothing more. When Jacobs tested his findings against measurements from the McAdoo tunnels, there was no evidence that the silt was allowing any settlement. But the nagging question of whether or not to use the screw piles was a complex and difficult one. The risks of not using them were obvious, but there were some risks inherent in using them, too. Not surprisingly, there was never unanimity among the engineers, but a decision to forgo the use of the piling was finally made by senior Pennsylvania management in 1907, and it was never installed.[25]

It would be almost another three years before the whole first phase of Alexander Cassatt's bold plan for the Pennsylvania would be in place. By the middle of August 1910 the tunneling had been completed, the electrification and signal systems had been installed and tested, the tracks had been put in place, and the new steel cars and electric locomotives had been started up. The grand New York structure that would be called Pennsylvania Station had been completed, scrubbed, and polished to perfection and admired by virtually everyone who saw it. For the first time, Pennsylvania Railroad trains would travel on their own tracks from Newark, across the Hackensack Meadows, through the Bergen Hill Tunnel, under the Hudson, into Penn Station, and on to Long Island via the Pennsylvania's own East River tunnels.

Between McAdoo's tunnels and the Pennsylvania's, the Hudson had finally been crossed at New York, in no uncertain terms. Everything had been built well and was there to stay. But history wasn't kind to the companies that had made it all happen.

McAdoo's Hudson and Manhattan Railroad would prosper until about the late 1920s, when competition from automobiles (intensified by new vehicular tunnels) began to drive down ridership and revenues. The railroad managed to hold its own during the war years, but afterward labor disputes, strikes, and struggles within the Hudson and Manhattan Company's own management combined with postwar inflation to weaken it further. By 1954 the company had been driven into bankruptcy.[26] In 1962, the Port Authority of New York acquired what remained of its corporate structure and integrated it into the Port Authority Trans-Hudson system (PATH), displacing and demolishing the Hudson Terminal office complex and building in its place the ultimately ill-fated World Trade Center.[27]

The Pennsylvania Railroad, once the largest publicly held corporation in America, didn't fare any better. Relying for much of its twentieth-century business on the Northeast's relatively short passenger and freight hauls, it saw the first of its of dramatic reductions in revenue as soon as the implications of the Federal Highway Act of 1956 began to play out. Railroads simply could not compete with the growing network of subsidized highways. Burgeoning automobile travel and then air travel exacerbated the Pennsylvania's problems, and the first public indication that the end might really be in sight came in 1963, when the railroad sold off Pennsylvania Station in New York and watched that glorious building succumb to the wrecker's ball. By 1968 the directors had decided that merging the railroad's declining assets with those of its long-time rival, the New York Central, might offer the last, best chance for survival. The two once-great companies joined forces in what was billed as the largest merger in railroad history, but it was to no avail. In June 1970 the Penn Central Corporation became bankrupt, and its properties, including rail lines and equipment that had been almost a century and a half in the making, were divided between Conrail and Amtrak.[28]

CHAPTER 6

The Bear Mountain Bridge

By late 1910, when the trains of the Pennsylvania Railroad began carrying passengers and freight between the company's splendid new terminal on Manhattan Island and such distant places as Baltimore, New Orleans, Chicago, St. Louis, and beyond, New York City had pretty much emerged from its relatively brief adolescence to enter its young adulthood.

In 1883 the city had been connected to Brooklyn by what was in 1910 still the longest suspension bridge in the world, and a merger with Brooklyn in 1898 increased New York's population to almost 3.5 million. It was then the most populous city in the hemisphere and would in a little more than a dozen years overtake London as the world's most populous city, as well as its financial capital. The nineteenth century had seen the beginning of electric power and telephones in much of the city, as well as the expansion of its urban infrastructure to include sophisticated water and sewer systems, an orderly network of mostly paved streets, and monumental public works that included Central Park, the Metropolitan Museum of Art, Carnegie Hall, and the Metropolitan Opera House. Effective if often corrupt city administrations had been in and out of office for years, producing five state governors in a single century. By early in the twentieth century, with immigration increasing at a rate that ensured both a rising demand for the city's goods and services and the manpower needed to produce them, elevated railways and a subway system were up and running. Within less than another generation, the city's increasing breadth and strength as a center of commerce, industry, and finance would confirm its place in the first tier of international cities, alongside London and Paris, a role that would have seemed sheer fantasy only a hundred years earlier.

It was in the beginning years of that extraordinary flowering of New York that Edward Henry Harriman was born, in 1848, the fourth of six children of a minister named Orlando Harriman and his wife, Cornelia. The couple could trace their lineages through several generations of propertied forebears, but although they had inherited some of their ancestors' tastes and a few of their pretensions, the Harrimans of 1848 simply didn't have the means to indulge in many of them. Orlando would soon leave his pulpit to head out alone for the American West, in search of a share of its riches, and would return empty-handed a year or so later to resume his ministerial duties in a modest parish in Jersey City.

Things didn't get much better. In 1860, when young Edward Henry Harriman was only twelve years old, his parents saw qualities in him that they felt justified a difficult decision to enroll him in the prestigious and expensive Trinity School across the river in New York. But even at that tender age Henry, as he was called by the family, appears to have been influenced by a combination of forces that would fifty years later be compelling factors in his becoming one of the richest men in the world: ambition and impatience. After just two years at Trinity School, he announced that he had had enough of formal education and was ready to make his way in the world of commerce. He made it clear that he wanted no part of merchandising or shipping, careers that were still sustaining some of his prospering relatives. Rather, he wanted to work on Wall Street. He got himself a job as a messenger with a small but well-established brokerage firm there, and within only a few years he had become the firm's managing clerk.[1]

His success was neither surprising nor much different from that of other ambitious boys who were able by virtue of their quick wits and good memories to separate themselves from their less-skilled competitors. Wall Street was still a relatively young place that required no extended period of apprenticeship and no long wait for aging superiors to die. Especially during and immediately after the Civil War, it was a marketplace where large sums of money could be made by traders profiting from the volatility of the times, and Harriman was an attentive student. By 1870 he was able to borrow $3,000 from a prospering uncle to acquire an exchange membership of his own, and he launched E. H. Harriman Company as an independent stock brokerage.[2]

He turned twenty-two that year, about the age at which Wall Street figures of a later period would be graduating from college and starting their careers, and his transition to the independent status of Wall Street principal effectively defined the beginning of his career. He moved quickly to cement relationships with prosperous family members, socializing with them and their friends and developing a substantial commission business in the process. He was a quick study, and it wouldn't be long before he was doing some profitable trading of his own, enhancing his income and his reputation as someone to reckon with on Wall Street. The fashionable town clubs of the era and Paul Smith's wilderness camps in the Adirondacks were places where the newly wealthy mixed with the establishment, and the old-timers were happy to welcome a bright young fellow like Harriman, who was clearly on his way up. He even found time to do some of the charity work that was more or less obligatory for an emerging financier.[3]

It was through his work with one of those charities that Harriman met Mary Averell about 1877. He'd been the prime figure in establishing New York's Tompkins Square Boys Club, which ministered to the needs of underprivileged boys, and Mary Averell of Ogdensburg, New York, had come to the city to visit relatives who had connections with the club. Just three years younger than Harriman, she and a brother were the only children of a prosperous Ogdensburg family that had banking and railroad interests. Well-educated by the day's standards and conservative in her outlook and habits, she offered a good balance to the sometimes frenetic intensity that had come to characterize Harriman's style. She was taller and broader than Harriman, who was five feet, four inches, and had a narrow build, and when she wore the high hats that were in fashion, the two of them made an odd-looking pair. But their relationship matured into an engagement and, in September 1879, into a marriage.[4]

It would be unkind and unjustified to suggest that the social or economic well-being of the Averell family was a significant factor in Harriman's decision to marry Mary, as there's certainly no compelling evidence of that. By 1879 he was a man of considerable substance in his own right, and the marriage has been described by persons who knew the couple well as having been one of exceptional warmth and strength.[5] But in fact Mary's father, William Averell, did prove to be extremely helpful to his son-in-law during those important early years when the young man was first beginning to think of railroad securities as an investor, not as a broker. At about the time of the wedding, Averell placed Harriman on the board of his Ogdensburg and Lake Champlain Railroad Company, and not long afterward steered him into similar positions with other railroads in northern and western New

York. These enterprises would become links in Harriman's later career in the world of railroads.

Probably the most important thing William Averell ever did for his new son-in-law was to bring Stuyvesant Fish onto the Ogdensburg and Lake Champlain board at the same time he brought in Harriman. A casual social acquaintance and occasional colleague of Harriman's, Fish was an intelligent, Princeton-educated Brahmin whose upper-class credentials and charm could (and would) open many doors for his friend. Handsome and popular, he was the son of Hamilton Fish, a former governor of New York and Ulysses S. Grant's former secretary of state. The younger Fish had only recently left Wall Street to join the board of the great Illinois Central Railroad. Soon after the two young men began working together on the affairs of the Ogdensburg and Lake Champlain, Fish induced Harriman to come along with him to see what he could contribute to solving some of the problems at the Illinois Central.[6]

It was a fortunate move for Harriman on a couple of levels. The post would be immensely profitable financially, placing him in a favored position for marketing (as a broker) the large security issues of a great railroad. On another level, the assignment thrust him into the heady world of railroad finance that would prove to be central to his own career. Harriman had begun to lose his taste for earning his living mainly by executing the trades of other investors, and he was drawn to the prospect of having a direct role in the world of railroads. Having learned a good deal about them on Wall Street, where railroad securities dominated nineteenth-century trading, he was especially knowledgeable about their operational subtleties, and he had some ideas of his own about how well-planned financial strategies could be as important to the success of a railroad as well-planned operating strategies. All those interests would be central to the close association he would have with Stuyvesant Fish over the years that followed.[7]

Harriman's broadening and often volatile commercial activity and his intense working style drove him relentlessly. After about six years, in 1886, by when he and Mary had two children (a third having died), he paused to look in another direction. During that year he purchased at auction a property of approximately 8,000 acres in the rural Ramapo Highlands of Orange County, New York, about ten miles west of the Hudson and about forty-five miles north of the family's East Fifty-fifth Street residence. The property was largely wilderness, covered by forests, laced with streams, abundantly populated by wildlife, crossed by mountain trails, and enriched by breathtaking views of that still unspoiled area of rural New York. The tract had a history that included several generations of iron mining and forging by the Parrott family, and its owners had been responsible for the historic Parrott Gun of the Civil War. The property included residences and outbuildings of the family, as well as other buildings occupied by tenant farmers and workers whose presence had been casually tolerated by the Parrotts for years. The place was called Arden, said to have been a family name, and it was to become a weekend and summer place where Harriman and his family could find refuge from the pressures of business.

Of course, Harriman was unlikely to approach even something as benign as a country retreat with any less vigor than he brought to all his other projects. Arden quickly began to evolve into another Harriman campaign. Within a year or so, he would engage superintendents and foremen to supervise maintenance and improvement of the place, and before long a substantial workforce of his own was blazing new trails, improving roads and drainage systems, and reclaiming some of the property's fallow land. By the turn of the century Harriman had built elegant new stables and become an important figure in the local horse breeding and racing community, and he had established Arden Farms, which would produce and sell dairy and other agricultural products that originated on the property. And, of course, he hadn't neglected the original residence, adding generously to its living and recreation spaces to ensure the comfort of what by then was becoming a family of seven and a cadre of resident servants.[8]

All this was being approached in the style of the very rich man that Harriman had by then become. Between his early and continuing successes on Wall Street and at the Illinois Central and greater ones that would follow at the Union Pacific and elsewhere, he would soon be one of the richest and most powerful men in the country. By 1902 he had acquired enough additional property at Arden to enlarge the original 8,000-acre tract to about 20,000 acres, a little more than thirty-one square miles. The rural outpost had become a thriving, populated community, complete with its own institutions, farms, shops, and utility systems, and Harriman was its benign if sometimes autocratic baron.

The work at Arden acquired a momentum of its own, and Harriman would soon engage the distinguished architectural firm of Carrère and Hastings, which had designed the New York Public Library, to crown the work with a splendid mansion on the scale of a royal castle. Between 1905 and 1909 a workforce that sometimes exceeded three hundred men worked to build one of those rambling palaces of native stone and timber and hand-carved wood that had become the hallmark of the American aristocrat. While Harriman's titanic and largely successful struggle to dominate American railroading continued, he managed simultaneously to build something that was at least as dear to him: a glorious, sprawling estate of his own in the splendid isolation of the Ramapo Highlands.[9]

But even estates as grand as Harriman's aren't invulnerable to the forces of outside events, and late in 1908 one of those events threatened the solitude of his beloved Arden. The superintendent of New York's prisons proposed that the state turn to nearby Bear Mountain as a virtually unlimited source of rock for building the region's public highways, and he convinced the Prison Commission to buy 700 acres of land there as the site of a new penitentiary to house the approximately two thousand prisoners who would do the work of drilling, blasting, and crushing the rock.

Having a prison as a close neighbor was something wholly unacceptable to Harriman, and he resolved quickly to resist the plan in any way he could. He petitioned Governor Charles Evans Hughes, offering to contribute 10,000 acres of his own land toward New York State's ambitious program for a new park in the Palisades and to contribute a million dollars toward its maintenance and care. All the governor had to do was to stop the prison plan. But between Hughes's busy schedule (and his apparently not altogether favorable view of Harriman's request) and Harriman's own impending trip to Europe, for what would prove to be an unsuccessful effort to recover his badly failing health, a meeting between the two men had to be deferred. It would never occur. Harriman returned from his trip in 1909, severely weakened by illness and exceedingly frail, and by August he was dead.[10]

His estate was enormous: the assets were estimated to be worth almost $200 million (several billion in twenty-first-century currency) and included control of about 75,000 miles of railroad. In his will Harriman left every bit of it to his widow, together with complete responsibility for managing it, and she quickly rose to the occasion. In December 1909, at a meeting of the new Palisades Interstate Park Commission, the details of an extremely generous bequest she was proposing to make to the park were spelled out, stipulating that it was contingent on matching cash gifts from others (including the states of New York and New Jersey) and establishing in clear language that the elimination of the prison plan was a non-negotiable quid pro quo. The formal proposal was transmitted to Governor Hughes, who guided it through the necessary legal steps. By the following year (with the considerable help of several Rockefellers, a Morgan, a Vanderbilt, and others) all the contingent financial conditions had been met or exceeded—and the prison plan was dead. In October 1910 eighteen-year-old W. Averell Harriman, just starting his freshman year at Yale, turned over the deed for the land and checks aggregating one million dollars on behalf of his mother.[11]

During the next decade or so, in the absence of the patriarch, things at Arden slowed a bit. While the Harriman children grew to adulthood, married, and prospered, Mary Harriman grew older, busying herself with the administration of the family fortune and quietly parceling parts of it out to what she considered to be deserving charities. Attendance at the Palisades Park grew slowly at first, while city people became gradually aware that the pleasures of outdoor recreation were so close and abundant, but it would soon accelerate.

During that same period, a change that was at once technological and cultural began to sweep across the country, and it was one that would have a radical impact on the Palisades. The motor car, which had first been produced commercially in Europe during the 1890s, had begun to have significant distribution in the United States in the early 1900s. Within fewer than twenty years the country was producing two million cars a year. By 1922 there were ten million cars registered in the country, and on a nice day in July it might have seemed to people waiting to cross the Hudson by ferry that most of them were headed for the Palisades.[12] On weekends, the waiting lines were usually more than a mile long, and people were reportedly sleeping in their cars overnight to ensure a place on the early morning boat. The ferry companies added equipment and accelerated their schedules, but those measures didn't help much. In 1922 the Dykman Street Ferry carried 325,000 cars across the Hudson, and the Nyack Ferry carried 107,000. Mary Harriman was said to be distressed about the situation, but there's no evidence that she offered to help or was aware of anything she could do about it.

That wasn't the case for an enterprising New York-based contractor named Frederick Tench, who saw conditions on the Hudson as ripe for a new bridge. Like the nineteenth-century bridge advocates who had preceded him, he saw the narrow passage between Anthony's Nose and Bear Mountain (just where a celebrated chain had been placed to obstruct the British fleet a century and a half earlier) as the ideal site for it. Born and raised in Canada, Tench had worked on steel for bridges in much of the United States, and in 1899 he had joined Edward Terry to establish Terry and Tench, a New York construction firm specializing in the erection of steel for buildings and bridges. Between 1899 and about 1922, the company had erected steel for such buildings as Grand Central Station and its neighbor, the high-rise Biltmore Hotel, and it had put up much of the steel for the Williamsburg Bridge and the Manhattan Bridge, on the East River. More recently, Terry and Tench had been erecting steel for railway viaducts and bridges across the Harlem River, some for the New York Central Railroad and some for the Brooklyn Rapid Transit Company. Its contracts were invariably for erecting only the steel, and there's no evidence that the company had ever taken a contract for building an entire bridge, but (like most such entrepreneurs) Tench apparently felt that his considerable experience in erecting steel would qualify him to build (and to estimate accurately the cost of building) a whole new bridge at Bear Mountain.[13]

He submitted his idea to Mary Harriman, and she passed it along to her younger son, Roland, who had taken over management responsibilities for the family's affairs. Twenty-seven years old and just five years out of Yale, Roland was intelligent, prudent, and sophisticated in matters of commerce. Although he wasn't immediately convinced that Tench's idea had merit, he invited him to develop details that could be further explored. That meant preliminary designs, cost estimates, and a construction plan, and Tench was of course eager to comply.[14]

During his work on the Harlem River bridges, Tench had become acquainted with an engineer named Howard Carter Baird, a Kentucky native who had spent the first seventeen years of his career working for the Phoenix Bridge Company of Phoenixville, Pennsylvania, the firm that had built the Hudson River bridges at Albany. A quiet bachelor, Baird had labored as a draftsman at Phoenix Bridge during the early years of his career there, studying civil engineering on his own and absorbing what he could of the knowledge of his superiors, and he had then continued on as a junior structural engineer himself at Phoenix. Baird was first formally recognized as a civil engineer in 1897, when the Engineers Club of Philadelphia accepted him as a member. A few years later, he left Phoenixville for Brooklyn, New York, to design bridges and other structures for the Brooklyn Rapid Transit Company, which was developing the city's elevated railway system. In 1908 he was hired away from the BRT by Boller and Hodge, an especially prestigious New York engineering firm that was designing one of the Harlem River crossings for which the Terry and Tench Company was erecting the steel, and it was on that job that a friendship between Baird and Fred Tench took root. During the next few years, both Boller and Hodge died, and in 1920 Baird established his own engineering practice. A little over a year later, he would be one of the first persons to register as a practicing civil engineer under the newly enacted registration law of New York State.[15]

In 1921 Baird and Tench were clearly a couple of fellows who needed one another. Tench needed a design for a bridge to cross the Hudson, and Baird needed a bridge commission to give his new career a start. After they joined forces, it didn't take Baird long to lay out an elegant, single (1,632-foot-long) suspension span across the Hudson at Bear Mountain, tied back to a couple of land-based rock anchorages and supported by two shore-based 350-foot-high steel towers. The prospective design came with enough details to enable Tench to produce a cost estimate. Baird was apparently unfazed by the fact that his design included the longest suspended span ever built, the previous records of the Brooklyn and Williamsburg bridges having been 1,596 feet and 1,600 feet, respectively. Later in 1921 Tench submitted Baird's design and his own detailed work plan to Roland Harriman, together with a construction estimate of about $3.7 million, and hoped for the best.

It didn't work. In June 1922 Roland advised him that, although he and his mother supported the idea of the bridge, neither of them was ready to accept the financial basis on which Tench proposed getting it built. Just what bothered Harriman is not clear, but there's some indication that Tench may have proposed that the work be done jointly by the Harriman family and Terry and Tench, with financial risks fully shared. The Harrimans, on the other hand, appeared to prefer an arm's-length construction contract with a fixed price for the bridge.[16]

Tench didn't interpret Roland Harriman's response as an absolute rejection, and he proved to be right. There were continuing discussions, and over the next few months Tench submitted a new proposal offering to do the work for a fixed price that was a little lower than his earlier estimate. With help from some of the Harrimans' well-connected friends, he managed to secure charters from the New York State Legislature and from the United States Congress authorizing construction of the bridge and the right to collect tolls on it. The Harrimans, mostly through the W. A. Harriman Company, which had been established on Wall Street by E. H. Harriman's brother William, sold about $4.5 million worth of bonds, mostly to the Harrimans themselves and to a few relatives and wealthy neighbors who shared the Harrimans' interest in the Palisades area. It was estimated that after a little early, temporary financing, the proceeds of the bond sales would be enough to build the bridge and to pay related expenses like land acquisition, engineering, financing, and supervision of construction. The toll-collecting capability of the corporation would provide for paying interest on the bonds and for redeeming them at a respectable profit to the investors after thirty-four years.[17]

On March 24, 1923, William Harriman's newly formed Bear Mountain Hudson River Bridge Company contracted with Terry and Tench to build the bridge and its approaches for $3,484,000, of which $500,000 would be paid in the form of the new corporation's bonds and the balance in cash.[18] Tench started work right away. His firm would do some of the fieldwork with forces in its own direct employ, but it would subcontract much of it, most notably the steel framing and cable work. Subcontracts were issued quickly, one for the structural steel to the McClintic Marshall Company (the giant Pennsylvania-based fabricator that would a few years later be absorbed by Bethlehem Steel Company), and one for the bridge cables and suspenders to John A. Roebling's Sons Company, the builders of the Brooklyn Bridge. By late spring 1923 drilling into solid granite for the four approximately 100-foot-long anchorage tunnels and for the two massive tower foundations had been started, and by fall the first concrete was being placed for the tower foundations. On October 22, just six months into the project, erection of the east tower had been started, and in spite of a cold New York State winter, things were going well. Both the east and the west towers were completed by April 1924, ready to support a system of wire rope cables that would carry a temporary footbridge (as a working walkway for erecting the main cables) across the river.[19]

In June, the critical task of spinning the bridge's permanent cables was started, and in about ten weeks Roebling's crews were able to put the two 18.5-inch-diameter cables in place, an extraordinary achievement. The company characterized its work on the Bear Mountain Bridge as a turning point in the methodology of cable spinning, mainly because it had for the first time successfully formed the wires into flat bands, eliminating the tedious cable compaction method used for earlier bridges and sharply reducing the time needed to put the cables in place. Almost forty years earlier, spinning the cables for the Brooklyn Bridge had taken sixteen months.[20]

Things appeared to be going well, but by about the middle of 1924 there were signs that the Terry and Tench Company was struggling for financial solvency. Whether the problem was the result of difficulties on another job, or whether Tench had simply underestimated the Bear Mountain job (and then reduced his price to conclude his negotiation with Roland Harriman) is not clear, as there had never been a competing cost estimate to compare with his. Whatever the reason, Terry and Tench defaulted on its contract and was reorganized as Tench Construction Company, with the bonding company taking financial responsibility for about a million dollars' worth of work that remained to be done. The default had to be a terribly demoralizing disaster for Fred Tench, who had a reputation for great integrity and enviable competence.[21] But construction continued without serious interruption, and the first of the bridge's stiffening trusses and horizontal bracing trusses arrived by barge around the first of September. They were in place within a little more than two months of their arrival, and Tench had substantially completed the paving of the bridge's 38-foot-wide, four-lane roadway as well.

Building the three and a half miles of road needed to connect the bridge's easterly landing to the nearest highway proved to be every bit as difficult as building the bridge itself, and it might well have been at the heart of Tench's financial difficulties. The road connection at the west end didn't pose any special problems, but the job of building a road between the easterly landing of the bridge and the National Guard camp on the road to Peekskill had been daunting from the beginning. Work on it had been started early in 1923, not long after the bridge foundations were begun, and it wasn't really finished until the early months of 1925. The twisting course of that narrow roadbed, some of it more than four hundred feet above the river below and at least a mile of it hovering over the New York Central's active rail lines, had to be carved from the steep rock face of the Manitou Hills. With only limited access for equipment, trucks, and personnel at the easterly end, more than 300,000 tons of material had to be excavated, two-thirds of it drilled and blasted from solid rock. Under the terms of Tench's unclassified construction contract, all responsibility for the huge cost of such work was his. Almost a century later, navigating that now smoothly paved and well-marked but still treacherous road remains a challenge for even the skilled driver of a modern vehicle.[22]

By Wednesday, 26 November 1924, work on the bridge and its approach roads was pretty much complete. Only minor tasks remained, including

FIGURE 9 The Bear Mountain Bridge. Photograph courtesy of the New York State Bridge Authority.

the placing of sidewalks and an asphalt topping on the bridge deck and some finishing touches on the approach road. A crowd estimated at several thousand persons drove the length of the almost complete approach road that day and then, four cars abreast, they drove across the bridge to watch the dedication ceremonies on the west bank. The West Point Military Band struck up "The Star Spangled Banner," Roland Harriman unveiled a plaque, and the great new bridge started its life.

At an elaborate luncheon that followed at the Bear Mountain Inn, about two hundred of those who had figured most prominently among the planners, designers, and builders of the bridge heard Roland Harriman publicly and individually thank as many of them as he could identify and extol the elegance of what they had wrought. More than a little appreciation was expressed by other dignitaries for the generosity of the Harriman family and for the job done by workers who had managed to complete such a substantial amount of construction in only twenty months. The next day was Thanksgiving Day. As soon as the bridge was opened for the regular flow of traffic, a few pedestrians paid fifteen cents to walk across, and several thousand motorists followed, paying a toll of eighty cents.

Of course, few things are perfect. Neither Frederick Tench nor Howard Baird, the contractor and the engineer, ever made it to the podium when Roland Harriman singled them out for special thanks and praise at the Bear Mountain Inn luncheon. Tench's old car had broken down on the bridge, obstructing the traffic behind it, and the two men who probably had more to do with building the bridge than anyone else were still pushing it across to the westerly landing when their names were called.[23] And not everyone shared the widely held view that the bridge was a thing of great beauty. *Scientific American* called it a "hideous artistic failure that obtrudes its ugliness upon one of the beauty spots of America," and the magazine's editors predicted that "future generations will ask what manner of people these were that stretched so crude a structure across our stately river"[24]

The bridge generated a profit in its first year, but during the twelve years that followed, the optimistic predictions of its sponsors were approached only four times. Traffic on the bridge simply didn't measure up to what had been expected. That poor showing wasn't entirely the fault of the planners, although they might have expected that tunnels and the new George Washington Bridge in Manhattan and the new Mid-Hudson Bridge at Poughkeepsie would divert some of their bridge's traffic. What they could not have been expected to predict was the Great Depression, which was clearly a factor in losses sustained during every year but one between 1932 and 1939.[25]

The original charter provided that the state could buy the bridge during certain periods between 1929 and 1949 for stipulated amounts and that the state would ultimately get the bridge free in 1954. There was no obligation to buy. Until the late thirties, the bridge's modest revenues continued to generate enough money to pay the interest on the bonds and to maintain a sinking fund that would provide for redeeming them, but ultimately the bondholders became unhappy and began to encourage the directors of the corporation to sell the bridge to the state for the best price they could get.

The state had reasons of its own for liking the idea, but it didn't want to pay too much. Once surveys and evaluations of the physical plant had been made and the financial condition and estimated future earnings of the bridge had been analyzed in painstaking detail, the venerable planner Robert Moses (who was then the chairman of the State Council of Parks) weighed in directly with the governor to urge that no more than $2.3 million be paid. There had been some talk about the generosity and noble purposes of the Harrimans to induce a sweetening of the sale price, but Moses stood firm, insisting that there was "nothing in the financial set-up of the Bear Mountain Bridge to show that it was, at any time, a genuinely philanthropic or altruistic enterprise." In August 1940 the state bought the bridge for $2,275,000, and the proceeds were sufficient to redeem at full value all the bonds that were still outstanding and to purchase the outstanding stock at a price that was a little higher than the original cost.[26]

Howard Baird went on in 1930 to design the soaring steel arch that carries the Taconic State Parkway across New York State's Croton Reservoir, a 750-foot span that didn't set any records but was a very impressive arch for its day. Mary Harriman died in 1932, Roland Harriman died in 1978, and W. Averell Harriman (after serving as ambassador to Russia and ambassador to Great Britain and after serving a term as governor of New York) died in 1986 at the age of ninety-four. Arden House was purchased by Columbia University, which used it as a conference center and retreat for a while before selling it to the Open Space Institute in 2007.

CHAPTER 7

The Holland and Lincoln Tunnels

When Frederick Tench and the Harrimans had those first conversations about the Bear Mountain Bridge, they weren't the only ones who had noticed that the automobile and the motor truck were transforming life in America, and they weren't even the first. In New York, where rail traffic in the McAdoo and Pennsylvania tunnels was already exceeding expectations and ferryboats were carrying about six million vehicles across the river every year, the car and the truck had already begun to make their impact known in no uncertain terms. Henry Ford had produced almost fifteen million cars by then, times were good, and the emerging concept of installment purchasing was bringing the automobile within reach of the working family. Trucks along the New York waterfront were lining up for hours, waiting to be loaded or unloaded. The federal government had passed a national highway funding bill, so new roads would soon encourage people to move out to the suburbs and drive to work, and Jersey farmers would send their products into New York by motor truck. All anyone needed, it seemed, was a way to get all those cars and trucks into and out of New York without having to rely on the ferries.

By 1913 New York and New Jersey were well past the question of whether or not they needed to build something. Instead, they were focusing on just what to build and where to build it. A bridge was clearly an attractive option, one that had been given a good deal of attention, but by 1913 the idea of a vehicle tunnel had been gathering support, too, encouraged by reports of successful tunnels in London, Glasgow, and Hamburg. To explore the matter, New York and New Jersey each established its own interstate bridge and tunnel commission.

The bridge boosters had been at their task for a long time and maintained a comfortable lead in the competition for public support, although that gap had begun to narrow a little as the arguments of the tunnel boosters gained a wider audience. Still, the bridge case was strong, and there was no scarcity of designs. That challenge had been around for a long time and had attracted the attention of more than a few prominent engineers.

As early as 1893, Union Bridge Company, builders of the much-admired Poughkeepsie Railroad Bridge, had been in the hunt, offering a design for a cantilever bridge to be located at Seventieth Street. The following year an exceptionally well-qualified board of engineers appointed by President Grover Cleveland produced a report that endorsed the feasibility of a cantilevered span as long as 3,100 feet. But the board ended up recommending a suspension bridge instead and proposed that it be built somewhere between Fifty-ninth Street and Sixty-ninth Street. Two years later George Morison, who had been a member of the board, presented his own design for such a bridge. An erudite, Harvard-educated bachelor who read Latin and Greek and had studied law before devoting himself to civil engineering, Morison had compelling credentials as the builder of railroad bridges across the Mississippi, the Missouri, and the Ohio. Any design he proposed would be difficult to ignore.[1]

Another bridge design presented in 1913 was the work of the eminent Gustav Lindenthal. Over the course of a long career, he had done a great deal of important work on bridges in New York and elsewhere without ever abandoning his passionate advocacy for the one he had first proposed to the Pennsylvania Railroad back in 1888. That structure, originally designed to bring all the railroads across the river at Twenty-third Street, had been compared by *Scientific American* with the pyramids of Egypt for its monumentality. Later, it would be modified for construction farther north, at Fifty-seventh Street, and would evolve into an even more ambitious concept that included eight sets of railroad tracks and two conveyor platforms on its lower level and four additional sets of tracks, six lanes of vehicle traffic, and two sidewalks on its upper level.

By the time his Hudson River bridge was being considered as an alternative to the vehicle tunnels, Lindenthal had recovered (as much as he would ever recover) from the Pennsylvania's 1902 decision to reject his bridge in favor of building its own tunnels, and he had busied himself ever since with other important work. For a time he had served as commissioner of bridges in New York City, and while the bridge and tunnel commissions pondered whether bridges or tunnels across the Hudson would be better, he was designing the Hell Gate Bridge to carry the Pennsylvania's trains across the East River.[2]

What might have been the best model for the commissions to see was a contemporaneous design of Boller, Hodge and Baird, the firm with which Howard Baird had been associated before he went off on his own to design the Bear Mountain Bridge. A suspension bridge, the Boller and Hodge structure was much lighter and more modest than Lindenthal's and included a suspended span of about 2,880 feet with a couple of approximately 1,000-foot-long side spans. Henry Hodge had estimated in 1912 that it could be built for about $30 million if located around Canal Street but that it was likely to cost about $40 million if built at Fifty-seventh Street. That latter figure eventually became the one used in comparisons with the cost of building a tunnel.[3]

Some of the designs considered by the commissions provided for both rail and automobile traffic, several chose locations south of Fifty-seventh Street, and all of them entered the city either at or south of the Seventies. The decision to provide for both rail and automobile traffic reflected a lingering uncertainty as to whether cars and trucks would simply become companions to the railroad or whether they would in fact dominate it to such an extent that no further rail capability would ever be needed in structures crossing the Hudson. As to location, the decision to keep the bridge south of the Seventies was likely to have been driven by the topography of adjacent shore areas and the depth to bedrock under the river. The shoreline topography was critical because the War Department required generous clearance below any bridge for river shipping, an easy matter farther north, where the adjacent banks were already high, but difficult and expensive where they were low. As to bedrock for foundation support under the river, its depth was known to increase sharply as the locations being considered moved north. Locating the bridge where the rock was down so far would have meant either very deep and expensive piers in the water or shallower, cheaper piers along the shorelines, where the rock was much closer to grade. Positioning the piers at or near the shoreline would have required an unsupported span between the towers that was still perceived by some engineers (despite the opinion of the Cleveland board) to be too long for contemporaneous technology. It would be another fourteen years before Othmar Ammann would dare to risk bridging the 3,500-foot span in one fell swoop with his George Washington Bridge at 179th Street.

The tunnel advocates had their work cut out for them, with all those bridge options and a slew of experts arrayed against them, but they were a determined lot and had a few compelling arguments of their own. They had no reason to consider the shipping clearance issue, and because they neither needed nor wanted high shoreline topography, they actually preferred the low-lying areas at the southerly end of New York and New Jersey, where they would have good access to what was then the center of commerce and industry in New York. They weren't troubled by deeply buried rock, either, as they preferred to bore their tunnel through the Hudson's silt than to drill and blast through its rock. There was an unexpected bonus, too, that wasn't recognized. The tunnel advocates benefited from poor cost estimating. They embraced the increasingly popular argument that a tunnel could be built for $11 million, a price that looked like a bargain when compared with Boller's estimate of $40 million for a bridge. In fact, neither of those estimates was anywhere near right. The tunnel would eventually cost $44 million, and although Boller's bridge was never built, the cost of later, similar bridges suggests that $60 million would probably have been a more realistic estimate.

Subaqueous tunneling was a much younger technology than bridge building, so the tunnel advocates couldn't match the bridge advocates in the number of experts they could bring to the debate. But they did have at least one distinguished professional in their corner. He was J. Vipond Davies, a native of Wales who had joined forces many years earlier with Charles Jacobs, the English-born engineer who had dominated tunnel work around New York since about 1890. Davies had prepared a design for a Hudson River tunnel that comprised two 30-foot-diameter tubes to be laid between Canal Street in New York and the Erie Railroad yards near Twelfth Street in Jersey City. That tunnel, and Davies's estimate that it could be built for only $11 million, became the tunnel model for both the New York and the New Jersey commissions.[4]

By the end of 1913 the two commissions had come to different conclusions. New Jersey favored a bridge at Fifty-seventh Street, and New York preferred a tunnel at Canal Street. Some commissioners in each camp recommended that both a bridge and a tunnel be built, but the financial means to carry out both projects wasn't within anyone's reach at that time. The serious work done by the two commissions was useful, though, because it vigorously dramatized the idea that something had to be done, and it gave needed credibility to a couple of solutions and locations.[5]

It would be almost another five years before matters would be resolved with any certainty. Comparisons of the two options continued to appear regularly in journals and the popular press, diminishing in frequency only when most of the arguments had already been heard several times and when reports of what became World War I began to drive bridge and tunnel news to the back pages. By 1918, bitterly cold winters and ice jams in the Hudson had several times brought New York to the brink of disaster, especially when the city found itself running out of coal for heating while piles of it sat within view on the Jersey piers. That frustration intensified pressures for establishing a new way of crossing the Hudson, and it clearly strengthened the arguments for a tunnel through which traffic could proceed despite ice, fog, and wind on the river. The commissioners themselves were by then beginning to run short of patience. Once the war ended, they resumed negotiations and soon agreed that it would be a tunnel between Canal Street and Jersey City, not an uptown bridge, that would link the two states.[6]

Each of the state commissions was understandably reluctant to abandon its independence, so in 1919 they compromised by moving into the same office in New York, allowing the members to manage the process together while continuing to look after the often divergent objectives of the states they represented. It was clear that they had done a good deal of effective preparatory groundwork for the tunnel, because the joint staff they established was able to move ahead with the business of design and construction with surprisingly little delay. One of its first acts was to appoint as chief engineer a thirty-six-year-old engineer named Clifford Holland, who'd already been working for a dozen years on the construction of tunnels under the East River. Originally hired fresh out of Harvard by New York's Board of Rapid Transit Commissioners, in 1906, Holland had started his career on the unusually difficult tunnel the board was already building to carry its new subway under the East River between New York's Battery and Joralemon Street, in Brooklyn, and he'd been engaged in building tunnels under the East River ever since.

Born in the Massachusetts town of Somerset in 1883, Holland had graduated from the prestigious Cambridge Latin School in 1901. Then he'd gone on to Harvard, where he combined a traditional liberal arts curriculum with a civil engineering program and earned degrees in both in five years. In July 1906, having passed a civil service examination he had taken before graduation, he secured the New York job and started work right away on the Battery-Joralemon Tunnel.[7] His work there was so exceptional that the journals he kept would become standard references for other engineers in the still young field of subaqueous tunneling.[8]

Not long after Holland started working on the East River tunnel, the newly established New York State Public Service Commission displaced the Rapid Transit Commission and assumed full responsibility for building, managing, and regulating construction of the subway system. It didn't take long for Holland to emerge as one of the new commission's stars.[9] For the next dozen years he worked for the Public Service Commission in positions of increasing authority on various East River tunnels, and by the end of 1918 he was responsible for an estimated $26 million worth of such work. A commission report characterized him as having "a breadth of information that few could equal."

A well-educated, self-assured engineer with clear leadership capability, Holland was a popular choice to spearhead the new Hudson tunnel. He had in 1908 married Massachusetts native Anna Coolidge Davenport, merging a couple of pedigreed New England families. Holland was descended from Roger Williams, a founder of Rhode Island, and his wife was a cousin of President Calvin Coolidge. The couple had three young daughters when Holland was appointed, and a fourth daughter would come along a couple of years later.

Before the end of his first few weeks in the new job, Holland had brought in two trusted colleagues to join him in the enterprise: Milton Freeman, a forty-eight-year-old civil engineer from upstate New York, and Ole Singstad, a Norwegian-born civil engineer almost exactly the same age as Holland. Not much is known of Freeman's earlier history, except that he had been working with Holland on some of the East River projects and was a seasoned construction man. Singstad had worked for several railroad companies, doing mostly bridge work before getting into tunnel work in 1909, when he moved to the Hudson and Manhattan Railroad. After a few years there, he had joined Holland on the East River work. When Jesse Snow, an engineer brought in by the joint commission, was added to the new staff, Holland had a three-person executive group that would serve him well in the years to come.

Not many details survive about Holland's personality, but there's plenty of evidence that he was a committed engineer and a decisive and effective manager. Most of his first year as chief engineer for the Hudson River Vehicular Tunnel was invested in reviewing and evaluating tunnel schemes that had been developed between 1913 and 1919, identifying what he felt had merit and then recruiting and managing a staff to design and build it. Eleven proposals made his first cut, all of them serious and well detailed, and Holland culled them with merciless thoroughness. By the end of his second review, he had pretty much defined his plan. Most of what he had discarded had failed for reasons of insufficient carrying capacity, poor constructibility, excessive time required for construction, failure to address one or more of the objectives of the commission, or probable cost. Several proposals were estimated by their designers to cost about $12 million to build, unsurprisingly close to the $11 million figure that the commission had itself carried forward from the Davies design of 1913 and made known to the candidates, but in Holland's report he rejected them all and established his own estimate of about $30 million. In fact, he was a good deal closer to the truth than anyone else, but he was still about $14 million too low.

And he hadn't allowed the prestige of advocates to influence his evaluations. One well-publicized design had come from the almost legendary civil engineer George Goethals, the West Point-educated general who had built the Panama Canal and served as governor of the Canal Zone. To buttress his presentation, Goethals had joined forces with John O'Rourke, the charismatic Irish-born civil engineer who had supervised construction of the Poughkeepsie Railroad Bridge a little more than thirty years earlier and had gone on from that job to a successful career as a contractor in New York. O'Rourke was no stranger to tunnel work or to Clifford Holland, having built at least one tunnel for Holland under the East River and having built the big Pennsylvania tunnels under the Hudson.

The Goethals plan was very ambitious, colossal in size and unprecedented in design. It was based on building a single 42-foot-diameter tube of discrete, precast concrete units and enclosing within it two wide levels of roadway, one above the other. Goethals and O'Rourke were in their mid-sixties at the time, and both may well have been positioning themselves for glorious conclusions to their professional careers. But the younger Holland would have none of it. He determined that their tunnel would never carry as many vehicles as they claimed, that the concrete units they proposed as a tunnel lining (in place of a monolithic steel, cast-iron, or concrete liner) would leak and be poorly suited to Hudson River conditions, and that the two-level design would require difficult and perhaps unsafe construction procedures, would take too long to build, and would cost too much.[10]

Another proposal destined for rejection came from Olaf Hoff, a distinguished tunneling engineer who only a few years earlier had completed work on a celebrated railroad tunnel under the Detroit River to connect Detroit, Michigan, with Windsor, Canada, and on another tunnel that would carry a New York subway line under the Harlem River. Hoff's proposal was based on using the open-trench method for building the tunnel, an approach he had used successfully on those two jobs in place of the more popular shield method advocated by Holland, in which a tunneling machine is urged forward through the riverbed by hydraulic jacks. In Hoff's trench method, deep trenches are excavated by dredging the river's bottom from moored vessels, and prefabricated sections of tunnel are lowered into place from barges and then backfilled. Holland argued that the trench method was poorly suited to the Hudson's conditions, mainly because the presence of moored vessels would be intolerably disruptive in a harbor as busy as New York's. Moreover, he argued, the silty soils of the Hudson's bed, unlike the clayey soils of Hoff's previous jobs, could not maintain steep channel banks and would generate a much larger volume of dredged material, increasing the cost. In addition, Holland said, tunnels that had used the trenching method had a poor accident record.[11]

Holland knew that almost everyone on his staff and virtually all the consultants he was interviewing had learned most of what they knew about tunnels from the McAdoo and Pennsylvania Railroad projects, either from working on their design or construction or from reading about them. But he was also aware that none of those tunnels had dealt with the problem of exhaust fumes from motor vehicles, as the trains they were designed to carry had all been electrically operated. Some information of limited value was available from a couple of vehicle tunnels that had been built under the Thames in London at about the turn of the century. Both the Blackwall Tunnel and the Rotherhithe Tunnel, which were being used by motor cars and trucks in 1919, relied entirely on whatever natural ventilation could be produced without the benefit of fans, and it was argued that no ill effects from exhaust fumes had been experienced in either. But Holland knew that the under-river length of each was no more than about one-third that of the proposed Hudson tunnel and that neither of the Thames tunnels was handling even as much as 5 percent of the traffic that was expected to travel through the Hudson River tunnel.[12]

Holland had to resolve the critical issue of ventilation before he could proceed with a complete design. He placed Ole Singstad in charge of the problem, and the two of them turned to the U.S. Bureau of Mines (and to Yale University and the University of Illinois) for exhaustive modeling and testing to determine just what should and could be done about exhaust fumes in the tunnel. The researchers confirmed an already widely held view that the only component that needed to be controlled was carbon monoxide, and that limiting it to four parts in ten thousand should ensure safety and comfort. To produce that level of dilution, they calculated, huge volumes of fresh air would have to be fed into the tunnel and equal volumes of vitiated air would have to be exhausted.[13]

The engineers' first idea was to bring the needed fresh air from a ventilating shaft at each shoreline and to induce it into a space below the tunnel's roadway. Fans would force the fresh air along the tunnel's length, distributing it in a carefully controlled way in the traveled area and retrieving it in a similarly controlled way at the tunnel's ceiling level. Finally, the vitiated air would be returned through the space above the ceiling to a shoreline exhaust shaft.

Much easier said than done. The volume of fresh air required to displace the fouled air with sufficient frequency was enormous, and the cost of power and equipment needed to move such a volume fast enough to achieve its purpose was excessive. They elected to solve the problem by bringing in the fresh air through a number of widely separated shafts, instead of only two, radically shortening the length of tunnel served by any one shaft and sharply reducing the volume and length of travel of air originating there. What evolved was a plan to build four ventilation shafts along the westbound tube's long route, and four more shafts along the eastbound tube's route, each of them complete with its own fans and control equipment. The shafts would serve as access routes for crews and equipment, too, and would be integrated into four permanent, aboveground (ten-story-high) ventilation structures that would enclose all the tunnel's air-handling equipment. Some of that early analysis had been done before Holland was appointed chief engineer, but ultimately it became the core of the planning and conceptual design work that would dominate his first year on his new job.[14]

Holland and his staff managed to resolve most of the major uncertainties by March 1920, when detailed design work could finally be started, but it wasn't easy. Some advocates of rejected designs were slow to be convinced, and General Goethals was one of them. When he successfully urged the commission to reconsider his ideas, Holland was obliged to review the proposal again and to give a second, more detailed defense of his decision to reject it. He restated and documented what he had already said about the plan's weaknesses, but this time he added a *coup de grâce*: Because the Goethals tube would be so large, he said, there were places along its route where the depth and weight of silt covering it would be insufficient to offset its buoyancy, and it could be expected to float. That ended consideration of the Goethals plan.[15]

During the early months of 1920, design work gained momentum and Holland's already difficult personal schedule intensified. The plan that finally evolved wasn't much different from what Holland had worked out in his review of prior schemes. It was to drive two parallel tubes, each approximately 30 feet in diameter, between the Canal Street site in Manhattan and the Twelfth Street site in Jersey City, with one tube carrying two lanes of westbound traffic, and the other carrying two lanes of eastbound traffic, both of them lined with concrete and equipped with sidewalks. They would be shield-driven, like the earlier Hudson tubes, and mechanically ventilated, and they would cost (it was still being said) about $30 million. Most of what was publicly announced in 1920 was true, but there's reason to believe that Holland and his staff already knew that their cost estimates were flawed. By that time they were well along in their ventilation studies, which showed the need for a vast amount of power and expensive equipment that hadn't been considered in the original cost estimates, and they had to be aware that wages and prices had increased significantly in the inflation that had followed the end of the war. In addition, there was evidence that plans for toll plazas and vehicular access at both portals were going to have to be dramatically revised and expanded, adding still more unanticipated cost. But like many planners who had preceded them and the many who would follow, Holland and his staff apparently decided to stay with their original figures until the project could gain some irreversible momentum.[16]

It would be fall of 1920 before the first dirt would be moved, but even then the project would be a long way from looking like a tunnel. The New York construction firm of Holbrook, Roberts and Collins had contracted to build the first access shafts, and its crews had begun relocating existing sewer and water piping in and near Canal Street, adjacent to the east portal, in preparation for building the shafts. That project was well along, almost a year later, when the carefully scheduled timing of bids for a second pair of shafts at the west portal was shattered by an unexpected irregularity in the bidding. Holbrook, Roberts and Collins was again the low bidder, but this time the firm had neglected to include with its bid the certified check required of all bidders as a security deposit to ensure performance. Bids for the second shaft contract had to be rejected and invited again, and this time (after the bids from the first round had been made public) the Holbrook firm was narrowly edged out by an insufficiently experienced competitor to whom Holland refused to award a contract. Over the objections of some of the commissioners, Holland stood his ground and rejected all the bids, electing instead to integrate construction of the western shafts into the main tunnel contract in 1922. He was said to be a man with a high level of self-confidence who was not inclined to back down when he felt the integrity of his tunnel might be at risk.[17]

Bidding for the main, double-tube tunnel contract, estimated by Holland's staff to cost about $20 million, wasn't as smooth a process as Holland would have liked, either. Bids were scheduled to be opened publicly early in February 1922, but only two firms showed up to submit prices. It was one of the largest fixed-price contracts in the city's history, and between the size of the job and the level of risk attached to any tunnel project, it's not surprising that many bidders avoided the job.[18] Holland realized that an award of this size based on only two bids might be tainted by suspicion of collusion, and he refused to open the bids, delaying the bidding for a few weeks in the hope of attracting another bidder or two. When the added time produced one additional bidder, Holland decided to take a chance and open the bids publicly, and he was glad he did. The lowest bid came from Booth and Flinn, a nationally prominent, Pittsburgh-based firm that had been in the heavy construction business since 1876 and was well known to Holland from earlier tunnel work. Its price was just under the estimate of $20 million, and the other two bids were a little higher.[19]

There was probably some celebrating in the engineering office after the bids were opened, but it's likely to have been restrained, as everyone knew it would be at least another eight months or so before any real tunnel boring would get underway. Much of the specially fabricated equipment and material that would be needed was only now about to be ordered. Most important were the caissons for the shafts, the shields for driving the tunnel itself, and the cast-iron frames that would be the tunnel's liner and permanent structure. They all required a good deal of engineering and production work, and they would all take time. The 400-ton tunneling shields would be made by the Merchant Shipbuilding Company in Chester, Pennsylvania; the huge caissons that would provide construction access and would become the structural frames of the shafts would come from nearby Staten Island

Shipbuilding Company, which would float them across the bay to the work site; and the almost 120,000 tons of curved, cast-iron ribs that, when bolted together, would provide the tunnel's watertight structural frame would come from the Bethlehem Steel Company's plant in Pittsburgh.[20]

Westbound boring for the north tube did in fact get underway in October 1922, and the work would proceed much as it had for the Pennsylvania Railroad's tunnels almost fifteen years earlier, except for the level of difficulty. The new tunnel at Canal Street was larger in diameter and would require more effort to drive it through the Hudson River silt. Its diameter of a little more than 29 feet was almost 25 percent greater than the Pennsylvania's, and that meant driving through a face area half again as big as the Pennsylvania's. To achieve that, each new shield would be pushed forward, 30 inches at a time, by thirty hydraulic rams with a combined force of 6,000 tons.

Once the western shafts were ready, additional shields began working their way eastward, starting in the south tube and following along later in the north tube. As in the Pennsylvania tunnels, the crews worked under air pressure to keep the headings tight and dry, and they followed every advance of the shield by immediately erecting the cast-iron rib segments that would be joined to one another to become the tunnel's cylindrical, enclosing structure. It would take fourteen of those rib segments to make one complete ring, and each segment weighed a little over a ton. The sandhogs bolted them together through matching flanges with bolts that weighed more than 10 pounds apiece, and it took a pair of strong men to handle the approximately 60-inch-long wrenches used to tighten them. Just about everything about this tunnel, from the size of its components to its degree of difficulty, was on a giant scale.[21]

Problems were abundant, but only occasionally unexpected. Driving the shields under local roads near the Canal Street shaft was especially difficult because the ground was laced with debris from abandoned piers and other old structures and because the thin covering of soil between the tunnel roof and the road above it posed a continuing threat of a blowout from the pressure of the tunnel's compressed air. In April 1923 the sandhogs went on strike for raises to about $8.50 per day (from $7.00) and for recognition of their union, but they settled after about a week for the raises alone, and they agreed to arbitrate future disputes.

As the shields moved out toward the center of the river, maintaining precisely correct level and direction was probably the job's nastiest technical problem, and it required constant and painstaking attention. Deviation was avoided mainly by adjusting the balance between the amount of excavated material that was allowed to ooze back into the heading itself through the shield's doors and the amount that was being pushed out of the way by the shield. But significant deviations, sometimes caused by changes in the nature of the material being penetrated, were bound to occur from time to time and were extremely difficult to correct.

It was all in a day's work for Clifford Holland, who was said to be an almost constant presence at the work site. In October 1923 the chairman of the New York State Vehicular Tunnel Commission announced that the tunnel workers had established a speed record in subaqueous tunneling by advancing 90 feet in a single, seven-day week, an average of a little more than 4 feet in each eight-hour shift. In the jargon of the sandhogs, that reflected almost two "shoves" of the shield during each shift, work that included advancing the shield and then erecting and bolting up the cast-iron sections that lay directly behind it.[22]

There was a kind of interim "judgment day" early in January 1924, when the tunnel commissions in New York and New Jersey issued their annual reports, complete with updated cost estimates for the work. The total projected cost was now about $44 million, an increase of about $14 million, but when it was all explained, the legislatures of both states appeared to take it in stride. For one thing, recent progress had been so good that it was hard to fault the people who had achieved it, and for another, there was evidence that most of those concerned had been fairly well prepared for the bad news.[23]

The crews labored on through 1924, and progress remained good. There was some satisfaction that although there were occasional occurrences of caisson disease ("the bends") caused by variations in air pressure, there had been no serious cases. Of course, that didn't mean that accidents could be completely avoided in such hazardous work, and thirteen men engaged in the work would die before everything was finished, some from accidents and some from illnesses contracted during the course of the job.

During the second week in October 1924, Holland himself began to feel some of the stress of his demanding work habits. With things going smoothly on the job and the "holing through" of the north tube anticipated for the twenty-ninth of the month, he went out to the Battle Creek Sanitarium in Michigan for some rest and recouperation. Some newspapers characterized his condition as a "nervous breakdown" and others described it as evidence of the "ravages of overwork." The sanitarium, founded in 1866, was the most famous health spa in the world at the time, and its natural, vegetarian approach to well-being attracted many pubic figures and celebrities in search of cures for everything from illness to fatigue. After appearing to have regained some of his strength, Holland suddenly died there of heart failure on the twenty-seventh of October. He was forty-one years old, and his death came just two days before the holing through of the tunnel.[24]

The shock of Holland's sudden death was pervasive at the work site and wherever people were involved in or simply interested in the work. Plans for celebrating the historic moment of holing through were canceled, and the commission decided to name the tunnel for Holland, making it one of the few public structures ever to have been given the name of its engineer. He was widely eulogized.

But the work had to continue, and it did. Milton Freeman, who had been Holland's senior assistant, took over and addressed the project with the same vigor that had characterized Holland's approach, restoring morale and momentum. Early in 1925, when preparations were being made for installation of the 14-inch concrete lining that would protect the interior face of the cast-iron framing, tragedy struck again: Freeman died of pneumonia. Managing such a project was clearly extraordinarily demanding, but having its two leaders struck down within a period of only six months seemed almost bizarre, and it had to be demoralizing to everyone involved.

Ole Singstad, who is said to have had a role in almost every tunnel project in New York between about 1920 and 1950, took over for Freeman. It is widely believed that although Clifford Holland had been the dominant personality in the conception and construction of the tunnel right up until his death, Singstad had been its principal designer, most especially of the novel ventilation system, which became the model for subsequent similar tunnels. His intimate and expert knowledge of every aspect of the work is often cited as the essential key to its successful completion. The two years of construction that remained would be punctuated by the recurrence of some of the earlier problems, but for the most part previously established solutions had simplified matters and things went well. By early 1927, there remained only the completion and fitting out of the ventilation buildings, installation of fire protection, telephone, security, and lighting systems, and construction of the plazas and new streets in a congested section of downtown New York. By November of that year Governors Moore of New Jersey and Smith of New York had pronounced the tunnel open for business.[25]

The Holland Tunnel was by any measure an overwhelming success. Admired as much for its effectiveness in meeting a profound public need as for its elegant technology, it demonstrated early along that it would pay its own way, too. Although the tunnel commissions had been compelled in the beginning to rely on the legislatures of New York and New Jersey for funding, it was now clear that tunnel traffic would exceed even the most optimistic predictions and that revenues would soon repay the states and generate surpluses. It was a success that almost everyone in New York had come to expect by 1927. In fact, for several years just before and after the tunnel's opening day, increasing concerns were heard that traffic might exceed capacity and that a second tunnel would be needed. As early as 1925, an investment group headed by a former state engineer sought authorization from the legislature in Albany to build a privately owned vehicle tunnel between lower Manhattan and New Jersey. By 1928 the Central Mercantile Association, the Lower Manhattan Industrial Association, the Fifth Avenue Association, and the press had joined a growing chorus for another tunnel. Almost every advocate favored a different plan, and the tunnel commissioners, justly proud of what they had done on the Holland Tunnel, were more than ready to step in, make the necessary decisions, and do it all over again.[26]

But something had changed. While the commissioners had been focused on building the Holland Tunnel, the governors of New York and New Jersey had joined forces with a group of committed visionaries who had been seeking for years to get the governors' signatures on an interstate compact that would authorize a single agency to develop, improve, and manage the complex port they shared. In 1921, they had succeeded, and the product of their agreement was the Port of New York Authority (later to be called the Port Authority of New York and New Jersey). What they created was an instrument of government that reflected an increasing public preference for the sometimes controversial idea of substituting control by appointed experts for control by politically vulnerable, elected officials in selected, especially complex aspects of governance. The new authority would have no taxing power, of course, but it would have the ability to design, build, and operate the port's transportation facilities, to charge fees for their use, and to issue bonds that would be repaid from revenues the facilities were expected to generate.

The Authority would have more than its share of growing pains during what remained of the 1920s, but by 1930 it would in every sense be a going concern with a strong recent past and a brilliant future. By then, with the indispensable help of New York's skillful governor, Franklin Roosevelt, it had taken as its not entirely willing bride the commission that had built the Holland Tunnel, and had emerged, not unexpectedly, as the thoroughly dominant partner, its operational philosophy and its mandate intact and its name on the door. By the time of the merger, the Port Authority had already built a couple of bridges connecting Staten Island with New Jersey, was well along on construction of both the Bayonne Bridge and the George Washington Bridge, and was about to become rich and more powerful by virtue of its newly acquired control of the bulging revenues generated by the Holland Tunnel. Early along, the Port Authority made the critical if controversial decision that revenues from all its present and future bridges and tunnels would be consolidated and pass through its own general fund so that all facilities could share the benefits of the agency's good credit. With its bridge-building program moving along well, the Port Authority would quickly turn its attention to construction of a second tunnel to New Jersey.[27]

It was decided that the second tunnel, which would (for a few years) be called the Midtown Hudson Tunnel, would connect a portal at Thirty-eighth Street in Manhattan with one across the river in Weehawken, New Jersey, and that it—like the Holland—would provide two lanes for traffic in each of two approximately 30-foot-diameter tubes. The Authority's engineers estimated a total project cost of about $95 million to build and equip the two-tube tunnel, including engineering, interest during construction, and contingencies, more than twice the cost of the first tunnel.[28]

Starting the process of building a second vehicle tunnel under the Hudson was reasonably expected to be a little easier than it had been ten years earlier. The experience gained on the Holland was all relatively recent, well documented, and accessible, but the task wasn't going to be as easy as expected. In 1931, in an environment of spiraling unemployment and failing businesses, there were clearly trouble spots ahead for any large, new project. The Port Authority faced one almost right away, when it found itself unable to market its bonds at tolerable interest rates, so it announced a delay of at least a year and suspended virtually all planning and design work until it could see a more favorable market for its bonds.[29]

But the agency didn't quit. Instead, it turned to the federal government, which was beginning to advance money for public works where the prospect of mitigating unemployment was clear. The Authority sought a federal loan of $75 million for the second tunnel and revealed in March 1933 that

Washington was prepared to make such a loan. But when the two parties got into the details, neither was happy with them, and during the months of negotiation that followed, a surprising compromise evolved. The Authority decided to build only one of the two tubes right away and to borrow only half the amount originally requested. It had decided to live with a single tube that would provide only one lane of traffic in each direction until a second tube became affordable. Once that compromise had been formalized, the Authority resumed engineering work and announced later in 1933 that it would be seeking bids for the south tube alone early in 1934.[30]

To chair the committee that would oversee the work, the Authority selected one of its own commissioners, a thoroughly experienced and supremely capable lawyer named Alexander Shamberg. His chief engineer would be the distinguished Othmar Ammann, who had not much earlier finished up his work as chief engineer for both the Bayonne Bridge and the George Washington Bridge. Ole Singstad had already done much of the required design work for a second tunnel, but by the time the project was ready to start, he had moved on to the New York Tunnel Authority, where he would become chief engineer and would head up design of the Queens-Midtown Tunnel. Ammann assembled a staff for managing construction that included Edward Stearns as assistant chief engineer, Ralph Smillie as chief of design, and Charles Gleim as chief of construction, all of them veterans of either the Holland Tunnel project or one of the bridge projects.[31]

There wasn't much good news anywhere in the winter of 1934, but the opening of bids for the first tube of the second tunnel is certain to have brought some cheer to Port Authority headquarters. The length of the single south tube was to be a little less than half the combined lengths of the two tubes of the Holland Tunnel, and the nature of the work to be done was virtually the same. As at the Holland Tunnel, the Midtown Hudson Tunnel's tube would be driven by the shield method, its structure would be an approximately 30-foot-diameter tube comprising 30-inch-wide cast-iron rings and lined with concrete, and its interior would be subdivided to allow for the distribution of ventilation air. Predicting the probable cost of this new tube should have been easy in 1934. Extrapolation of the twelve-year-old figures from the Holland, after adjustment for the shorter length resulting from the decision to build only one tube and for minor changes in specification, produced an expected low bid of about $8.7 million for construction of the Midtown's south tube. But it didn't work out that way. When the bids were opened, it was found that Mason and Hanger, an old and well-established firm that would soon be working on construction of the Grand Coulee Dam, had submitted a bid of only about $6.5 million, almost $1 million lower than the second bidder's price, and almost $2.3 million less than the fifth (highest) bidder's. The Great Depression was causing pain almost everywhere, but on that February morning in 1934 it appeared (to the tunnel's engineers) to be a cause for celebration.[32]

In good times, people are sometimes surprised that anyone seeks the difficult and dangerous jobs that require working in the surreal depths of mines and tunnels, but during the depression year of 1934 there was no such surprise. Jobs were almost impossible to find, and at least three or four applicants for tunnel jobs were regularly rejected for every man who was hired. As soon as the shafts had been prepared and the first shields had been put in place for the new tube, dozens of sandhogs began passing through the airlocks every day to work for limited periods in the dangerous depths of the tunnel. The work eventually settled into a rhythm of its own, with occasional exceptional progress. One crew was able to advance its heading a record 250 feet in a single, seven-day week, almost doubling the best day on the Holland job. But most weeks were much less successful, and it would be August 1935 before the shields being driven toward each other from opposite sides of the river would meet near the middle.[33]

Things soon began looking up a bit at the Port Authority's office. Once there was a sign of improvement in the financial markets, the Authority was able to sell some of its bonds, repay most of what it had borrowed from the government, and get the second tunnel underway. It was an event that drew the special notice of Harold Ickes, President Roosevelt's secretary of the interior, who publicized it widely as a good example of how the public works loan program was supposed to work.[34]

Whether or not reasonable progress or an occasional brightening of the financial skies had any bearing on the Authority's thinking about starting work on the second (north) tube of the second tunnel isn't clear, but in 1936 there were signs that the agency might soon go ahead with it. The press and local groups were saying publicly that the north tube would be needed as soon as the south tube opened; all seemed agreed that maintaining two-way traffic in the tunnel with only one lane operating in each direction could be problematic. There was much said about prudent management itself being a strong argument for shortening the period of construction by finishing the new tunnel sooner rather than later. Not surprisingly, the Authority wasn't hard to convince, and it reacted by inviting bids for the north tube during the closing weeks of 1936, scheduling a February 1937 date for opening them.

The response was disappointing. The only bid received came from Mason and Hanger, the company that was building the south tube. At about $8.8 million, it exceeded by almost 35 percent the price the contractor was being paid for the first, slightly longer tube, a price that had been established only three years earlier in a deflating economy. It seemed likely that the contractor's mere presence on the job and the advantage that attached to its having millions of dollars' worth of equipment already deployed there had discouraged everyone else from bidding. There was some concern that Mason and Hanger, which "had left almost a million dollars on the table" in the 1934 bidding, might have played an active role in discouraging other bidders, too, but there was no solid evidence of that. It was a bad situation, and the management of the Port Authority thought long and hard about the prudence of awarding such a large contract on the basis of only one bid. In the end the Authority apparently felt that there was little likelihood of generating effective competition while the south tube contractor was on the job and that the importance of maintaining a continuing source of employment and providing the best and most complete facility as soon as possible transcended other factors. It awarded the job to Mason and Hanger, which was able to move ahead quickly.[35]

A few months later, the Authority addressed the less profound but increasingly irritating problem of confusion between its own Hudson Midtown Tunnel and the City of New York's similarly named Queens Midtown Tunnel under the East River, on which construction was about to start. Both were beginning to be called the Midtown Tunnel. The Port Authority deferred to the city by formally changing the name of its tunnel to the Lincoln Tunnel, in honor of the country's sixteenth president.[36]

Orders for cast-iron and steel structural sections and for other critical components of the north tube had already been placed by the Authority when the contract for driving the tunnel itself was awarded, and because the contractor had long since deployed staff and equipment at the site, it was easy and natural for work on the north tube to get underway almost as soon as contracts were signed. But most of the attention in those early months of 1937 was still focused on the south tube, where no effort was being spared to meet the goal of opening the first tube before the end of the year. The concrete work inside it had been completed, and now a glass tile lining was being applied, wiring and piping were being installed, and the ventilation system was being finished up and tested. The south tube of what was now being called the Lincoln Tunnel was being made ready for a gala opening in December 1937.[37]

The goal was reached, with a week or two to spare. Toward the end of November the Port Authority began escorting schoolchildren and others on tours through the substantially completed tunnel, newly hired tunnel police were practicing their routines, and hundreds of photographs were being taken. On 21 December 1937 thousands of spectators and several military bands were on hand to hear former president Herbert Hoover and the governors of New York and New Jersey praise and dedicate the new tunnel. The next morning, cars and trucks started driving through, and by the first week in 1938 things had begun to settle into a routine, allowing a chance to analyze the new traffic patterns.[38]

In fact, despite the hoopla, things hadn't gone well at all. During the tunnel's first nine days of operation, which included the usually heavy travel of Christmas Day, an average of only about 7,600 vehicles per day had passed through it. The Holland had averaged about 11,700 vehicles per day per tube during its first year, and it had carried more than twice that many cars per tube on its first day. The Lincoln's figures were well below expectations, and they would get a good deal worse during the months that followed. One explanation blamed the shortfall in traffic on New Jersey's failure to complete connecting road systems at the western terminus of the new tube, but even that failure didn't seem sufficient to explain the problem. The Great Depression was still taking a heavy toll in 1938, and it may be that the optimism that is almost always basic to the outlook of planners and designers temporarily concealed that stark truth from them.

Nothing could be done about the completed south tube, but there was a good deal of rethinking about the Authority's decision to proceed with the north tube. Later in 1938 it announced that the north tube's completion would be deferred until traffic warranted it. Because the cost penalties for canceling contracts already in progress for the north tube were so severe, the Authority did not actually stop construction until October 1939, allowing most of the major work to be completed and leaving approximately $12 million worth of interior, mechanical, and electrical work, together with the construction of the approaches, to be done when the project was resumed.[39]

By early 1941, almost two years after the north tube had been sealed off, conditions for finishing the tunnel looked better. Road traffic had increased, the economy had shown some new strength, and New Jersey had significantly improved the network of roads that would feed traffic into the tunnel. Work on the north tube was resumed at the end of April, with a goal of finishing by the middle of 1943, but the impact of World War II on the availability of manpower and materials would delay completion until February 1945.[40]

Once the war ended, everything in New York—including its traffic—grew by leaps and bounds. In 1951 the Port Authority found itself long past any regrets about its decision to build the north tube. Its new concern was that both tunnels were being taxed to capacity and that additional construction was needed. Traffic patterns in the Lincoln Tunnel convinced its engineers that a third tube with two lanes would make sense, but they developed a novel plan for optimizing its use. In their plan, the third tube would sometimes carry two lanes of eastbound traffic, sometimes two lanes of westbound traffic and sometimes one lane of each. It would all depend on what was required and when. They positioned the third tube south of the two tubes that were already in place: What had been the south tube became the center tube, where traffic could be made to flow in either direction, and the added tube became the new south tube.

In October 1953 Mason and Hanger, ideally positioned to continue at the site, teamed up with two other contractors, the Arthur A. Johnson Corporation and MacLean-Grove and Company, to submit the winning bid to build the third tube of the Lincoln Tunnel. This time their contract would be for a little over $17 million and their goal was to finish by fall 1956, but they needed until May 1957 to do it. The total cost of the third Lincoln tube, once all its other costs are added to the amount of the Mason-Johnson-MacLean contract, was about $100 million, a little more than the cost of the Lincoln's first two tubes combined and about two and a half times the cost of the original, two-tube Holland Tunnel.[41]

During the 1990s, the Lincoln Tunnel is estimated to have been carrying about 43,000 vehicles per day through each of its three tubes, and the Holland Tunnel is estimated to have been carrying about 50,000 vehicles per day through each of its two tubes.

CHAPTER 8

The George Washington Bridge

THE HOLLAND AND LINCOLN TUNNELS were a resounding success. Except for problems that are almost inevitable when large and complex projects are built below ground, construction went well, and most of it was completed pretty much as scheduled. The costs proved to be a good deal higher than anticipated, vanquishing once and for all the idea that tunnels would always be cheaper than bridges. But the public demonstrated its usual tolerance for cost overruns in large public works and moved on. Once the effects of the Great Depression and World War II had waned, increasing traffic would justify the planners' projections (and then some), the tunnels would perform as promised, and almost everyone would be satisfied.

But even before much work on the tunnels had been done, serious talk about a possible additional crossing farther uptown was starting to be heard.[1] There was good support for the idea, but whether a bridge or a tunnel should be built was something that was going to require a good deal more study.

The valley's geology, which begins to change as the site is moved north, did not favor the construction of a tunnel north of about Midtown. Beyond that, the river had carved a path through a submerged mountain of granite that would be entirely inhospitable to tunnel construction but well suited to the support of tower foundations. In addition, the high banks that border the river up there are ideal for the approaches to a suspension bridge that must maintain plenty of vertical clearnace for river traffic.

That was good news for the bridge advocates, but it wasn't unalloyed. They had to be pleased (but hardly surprised) to know that maintaining the required clearance between the bottom of the bridge and the surface of the water would be much easier in a northerly location, and that the task of carrying traffic on and off such an elevated bridge would be easier, too, where the adjacent terrain was high. But there was still the issue of the War Department's prohibition against piers in the water. Building without piers in the water would mean a clear span of about 3,500 feet, and in the 1920s that was something that most engineers were not yet ready to try.

The bridge favorable geology of a more northerly site wasn't the only thing that was new for planners thinking about an uptown crossing in the 1920s. Another was the increasing ability of American industry to produce and sell large numbers of affordable trucks and passenger cars. Their drivers were already waiting in long lines for a chance to board a ferry, and as travelers and shippers moved away from railroads and toward automobiles and trucks, things could only get more difficult for the ferries. That historic shift in commercial and personal preferences would reduce the need to design bridges to carry the heavy loads of locomotives and freight cars and open the potential for lighter, less expensive bridges.[2]

Such geological and commercial realities certainly enhanced the case for a bridge, but other factors helped, too. Civic pride and an essentially optimistic outlook characterized much of popular thinking in the ("roaring") twenties, and it's likely that nothing less than a bridge would have satisfied the popular yearning for a monument that would give form to that spirit. While the tunnel was acknowledged to be a modern marvel, too, it was seen by some as lacking the majesty of a great bridge.

Many of the engineers who had figured in prior iterations of the long campaign to build a Hudson River bridge, either as designers or as critics, had died or simply become unavailable for the commission to design one in the twenties. Louis G. F. Bouscaren, who had not much earlier designed the longest truss span in the world, was dead, and so were Lefferts Buck and Thomas Clarke, designers, respectively, of the Williamsburg Bridge and the Poughkeepsie Railroad Bridge. George Morison and C. W. Raymond were both gone, and so was Theodore Cooper, designer of the ill-fated Quebec Bridge. Both Alfred Boller and Henry Hodge, whose bridge design had been the model that finally lost out to the Holland Tunnel, had both died. Howard Baird was immersed in his work on the Bear Mountain Bridge, and Ralph Modjeski was busy with a new bridge across the Delaware and preparing to deal with one in Poughkeepsie. There were of course other exceptionally well-prepared engineers who would have relished the chance to design a bridge on the Hudson, but few of them could match qualifications with the especially well-regarded and persevering Gustav Lindenthal.

When the idea of building a bridge across the Hudson in New York began to regain favor during the early 1920s, there were at least a few people around who could remember when the now well-established Lindenthal had first advocated his grand Twenty-third Street design more than thirty years earlier, and how the Pennsylvania Railroad had at first endorsed and vigorously supported it and then abandoned it. The disappointed Lindenthal had been left high and dry, and had placed his drawings in storage, along with the federal charter that he had wisely secured.[3]

After that, he had served the railroad on an advisory board for a while, in a position that might have been a well-deserved consolation prize, but in 1901 he had moved to a more fulfilling job as commissioner of bridges for the City of New York. The new position had the impermanence of any political appointment, but it was one in which Lindenthal was able effectively, if sometimes controversially, to oversee (and to varying degrees, participate in) the design and construction of three important and historic bridges across the East River: the Williamsburg, the Manhattan, and the Blackwell's Island (later called the Queensborough).

About 1905, after a new mayor had replaced him as commissioner, Lindenthal's fortunes took a more important turn for the better. Samuel Rea, his original patron at the Pennsylvania and still an advocate for his Hudson River bridge, had been given senior responsibility for the substantial job of connecting the tracks that carried the Pennsylvania's trains onto Long Island with the tracks of the New York, New Haven and Hartford Railroad. That was a long-sought linkage that would finally connect the Pennsylvania to the rich markets of New England. The gap between the ends of those systems was only a few miles long, but closing it required building about three miles of elevated structure across some poor ground, close to some sensitive buildings, and across sections of the East River that surrounded Ward's Island and Randall's Island. Probably the most challenging part of the whole task was building a major railroad bridge about a thousand feet long across a treacherous East River channel called Hell Gate without obstructing river traffic during or after construction.

Rea, who would soon become president of the Pennsylvania, awarded Lindenthal the commission to design the whole project. It was an especially attractive opportunity for a man whose work in recent years had been well suited to the needs of an independent, sole practitioner, but not to those of an engineer whose ambition was to design major bridges. Not irrelevant was the fact that the Hell Gate project was a very large one that would require a big staff and take a long time to complete. It offered Lindenthal a real chance to build the kind of engineering organization he would need for continuing later with other large bridge work, perhaps including the Hudson River bridge that was never far from his thinking.

There was no scarcity of problems to be solved on the Hell Gate project. Some of the especially difficult ones were attributable to the challenges of the river itself and the proximity of hospital and prison buildings, some to Lindenthal's determination to ensure that the widest possible variety of solutions be considered, and probably a few to his genuine wish to make an attractive aesthetic statement with the bridge's towers and its structure. It would be seven years before the conceptual phase of the design was finished and the development of working drawings was sufficiently advanced to allow for the start of construction.[4]

By then, Lindenthal had begun to think seriously about engaging someone to relieve him of some of the burden of the Hell Gate work. A few years earlier, at a social event in Philadelphia, he had been introduced by a colleague named Frederic Kunz to a promising candidate: a young, Swiss-born and Europe-educated bridge engineer named Othmar Ammann. Kunz was himself a well-regarded bridge engineer of considerable reputation, a partner of the equally distinguished bridge engineer C. C. Schneider, and young Ammann had been working with or for him and his partner in variety of professional relationships for a few years. Kunz had been sufficiently impressed to entrust Ammann with some sensitive design responsibilities and with the writing of limited parts of a textbook he was preparing with Schneider.

That first meeting between Lindenthal and Ammann was fortuitous. The two men and their wives met several times afterward, forming a friendship, and when Lindenthal learned in 1912 that Ammann was considering a career change, he invited him to become his senior assistant on the Hell Gate project, which was about to enter its construction phase. In fact, Ammann had been considering returning to Europe at about that time, but when Lindenthal sweetened his first offer, the younger man accepted.[5]

It's doubtful that the two men had much in common beyond their towering intelligence, their extraordinary articulateness in a second language, and their passion for building bridges. Ammann was thirty-three years old in 1912, a slight, clean-shaven, introspective man who was by most accounts unlikely to socialize much and whose voice wasn't always easy to hear. Whatever young Ammann may have lacked in social assertiveness, Lindenthal appears to have had in abundance. Sixty-two in 1912, he was tall and beefy, with a full beard, described by contemporaries as extroverted and certainly gregarious, and he spoke with a voice loud enough to dominate most conversations. Although the initial relationship between the two men seemed genuinely warm and mutually respectful, it would eventually rupture.

Ammann had been in the United States for about eight years by the time he took the job with Lindenthal, and a casual look at the meandering course of his earlier career might have suggested an undisciplined randomness. But just the opposite was the case. He had come to the country with a first-rate engineering education and a compelling interest in bridges that soon focused on very long spans. The evidence suggests that early along he had set out to learn everything he could about their design, fabrication, and construction. His objective, it now seems clear, was to acquire and combine such broad knowledge with the education he had received in Switzerland. Resisting the temptation to find and settle into an economically secure niche in a traditional engineering office, he had accepted varied employment opportunities during those early years in a range of settings calculated to broaden and enrich his knowledge of big bridges. And he acquitted himself well in no fewer than six different positions during that eight-year period: three with prominent engineers who specialized in the design of long-span bridges and three with big steel companies (two were nonconsecutive positions at the same company) that fabricated and built them. The engineers were an elite lot: Joseph Mayer in New York; Ralph Modjeski in Chicago; and Kunz. The fabricators were the Pennsylvania Steel Company near Harrisburg, largely controlled by the Pennsylvania Railroad, and the McClintic Marshall Steel Company in Pottstown, Pennsylvania, another giant that only a few years later would become the flagship division of the Bethlehem Steel Company. All of these experiences became part of a postgraduate education that took young Ammann from the drafting room to the fabricating shop and the construction site, and not a bit of this time appears to have been wasted.[6]

Ammann's work at Hell Gate became another phase of his broadening education. According to his diary, he was designated assistant chief engineer, reporting directly to Lindenthal, and was given senior responsibility for all office, field, and inspection work on the Hell Gate job. With construction starting, the design work must have been well advanced, so although Ammann's responsibilities probably included some design work, he was more likely to have concentrated on ensuring that fieldwork was being properly done and on addressing the myriad issues that inevitably surface when a complex project is executed in the field. Erecting the great Hell Gate arch itself, the central and most recognizable element of the project, was especially challenging and was said to have been masterfully managed by Ammann.

By late 1916, when the work was approaching completion and had begun to wind down, preparations for American involvement in World War I began to absorb the energies and resources of the country. Plans for new bridges and other nonmilitary construction projects were shelved, and any hope Lindenthal might have harbored that his staff would move seamlessly into work on a Hudson River project faded quickly. He already had a few other consulting commissions, providing some work for himself and Ammann, but those jobs would soon be finished, leaving Lindenthal's once-crowded drafting room distressingly empty.

Ammann had by then written and would soon present to the American Society of Civil Engineers a well-crafted account of the Hell Gate work that was so thorough and elegantly written that it was awarded the society's annual Rowland Prize for the best paper on an engineering achievement of the preceding year. Sixteen years earlier, the first Rowland Prize had gone to Lindenthal.

An old-school practitioner not likely to keep anyone on his payroll without income-producing work to cover his salary, Lindenthal worried about the possibility of losing his talented assistant, and he may even have felt some reluctance to abandon a loyal and productive employee to the vagaries of an often volatile marketplace. He looked to New Jersey, where he had for some years held a financial interest in a clay mine and pottery business that was now showing signs of failing. When it became understood that a new manager might be able to salvage the business, Lindenthal recommended Ammann for the job and urged him to take it. A far cry from bridge building, the move was clearly intended as a stopgap that would tide the younger man over until conditions improved, and Ammann agreed to make the change. Not surprisingly, the business proved to be one of those situations in which the uncluttered thinking of a good engineer was just what was needed, and by 1920 Ammann had restored the company to profitability and returned to his old job as Lindenthal's assistant.[7]

It was 1920, and a conclusion was finally being reached in the long debate between the advocates of building a vehicle tunnel between New York and New Jersey and the advocates of building a bridge. A variety of bridge schemes had been explored, including Lindenthal's, along with a number of tunnel schemes, and the bistate commission had decided in favor of a tunnel. Clifford Holland had been hired, design work was underway, and later in the year fieldwork on what would eventually be called the Holland Tunnel would start.

But most people close to the situation recognized that the tunnel was likely to be only the first of several crossings. Certainly it was clear to Ammann and Lindenthal that the decision to move ahead with a tunnel didn't necessarily mean the end of their bridge. Ammann, at forty-one, saw construction of the Lindenthal bridge as a critical next step in his own promising career, and he was ready to do whatever he could to develop its design further and to encourage its construction. For Lindenthal, at seventy, there was even more at stake: This was his last chance to build the great bridge that had been at the center of his life since he had conceived it thirty years earlier. After years of effort and disappointment, and with the end of his own productive career in sight, he was not about to let the prize slip away without a struggle.

By the time Ammann returned from his New Jersey interlude, Lindenthal had done a good deal of additional design work on the bridge himself, and he was well into the critical business of rounding up the technical, financial, and political support he would need to get it built. During what remained of 1920 and 1921, while Ammann and a small staff of engineers worked on completing and refining the design work, Lindenthal was out rallying support. Between letters to journals and newspapers and speeches at public meetings, he was campaigning with the vigor of a politician running for office, describing the virtues of his bridge, announcing and defending his decision to move its location from the original Twenty-third Street site to Fifty-seventh Street, and arguing for the economic advantages of bridges over tunnels.

Not surprisingly, Lindenthal was now proposing changes to his earlier design. By 1920 he had been pretty much forced to move his bridge north by a few miles, to Fifty-seventh Street, to avoid conflict with tunnel traffic. In addition, he was beginning to cede some ground to critics who argued that the volume of automobile and truck traffic would soon exceed the volume of rail traffic, and he was considering reducing or perhaps even entirely eliminating railroads from his planning, although that would later prove to be too difficult a pill for him to swallow. By 1921 the details of his redesigned bridge had begun to emerge, showing sixteen lanes of automobile and truck traffic and four sets of light rail tracks on one level and still showing ten sets of conventional railroad tracks on a second. It was still a vast project, and he had increased his estimate of cost to $100 million, a figure that didn't include such significant expenses as land acquisition, construction of approaches, or construction of the rail systems that would be required for connecting the new uptown site with downtown freight and passenger distribution centers. By later in the year it was becoming evident that the real cost of Lindenthal's revised scheme, including all those related items, would probably be closer to $200 million, and there were believable, higher estimates being heard, too.

What Lindenthal was proposing would be the most expensive bridge in history. Some critics argued that even the Fifty-seventh Street location was too far south, and there were other objections, too. By 1922 Ammann had begun to worry seriously about whether public resistance to Lindenthal's design might fuel a resurgence of support for more tunnels and bring an end to the bridge concept. Knowing how quickly a high cost alone could erode any support that Lindenthal had been able to marshal, Ammann began an earnest effort to induce him to make more radical concessions, urging that he eliminate provisions for rail traffic and reconsider design criteria that seemed unrealistic to the younger engineer. In addition, he proposed considering a site at or close to 179th Street, opposite Fort Lee, New Jersey.

Most of these suggestions simply went too far for the older man, and he reacted with unconcealed pique. He accused Ammann of timidity and short-sightedness, and early in 1923 the once solid friendship between the two men began to unravel. Although there may always have been a certain inevitability to a split, it doesn't appear to have become at all problematic until after Ammann returned from his New Jersey interlude and began to work on the Hudson River project. Only then did a fundamental difference of objectives enter the relationship. Contemporaneous, outside events beyond the control of the two men probably had some bearing on the environment in which a decision to separate developed. One of those events was the emergence of the Port of New York Authority as a potentially major factor in decisions about future bridge construction in the region, and the other was the election of George Silzer as governor of New Jersey.[8]

The Port Authority had been established in 1921 as an independent agency of New York and New Jersey, with a complex mandate to build, buy, and lease transportation facilities in an approximately 1,500-square-mile area that was defined as the Port of New York. The implications of that mandate were broad and subtle. The agency focused at first strictly on the movement of railroads and freight between the states, but some occasionally imaginative interpretations of its mission could be seen as a basis for a wider agenda. The Authority was already beginning to consider the construction of bridges, and within a few more years it would become the owner and controlling force in the destiny of the river's tunnels. As early as 1923, the Port Authority was beginning to be seen by some as the likely candidate for building and operating a bridge across the Hudson between New York and New Jersey.

As to George Silzer, who had become governor of New Jersey on the first of January 1923, he was a strong political figure who had over the years actively supported many rationally planned public works and had in his inaugural address expressed support for the idea of a Hudson River bridge. What makes Silzer noteworthy in the context of Ammann's 1923 break with Lindenthal is that the governor was a friend of both. Like Lindenthal, he had been a director of the clay mining and pottery company that Ammann managed from 1917 to 1920, and it's likely that Silzer and Ammann had become well acquainted during those years. The two men may well have discussed the virtues of a bridge across the Hudson and perhaps some of Ammann's own ideas about it.

Ammann is known to have had extensive contact with Silzer during the months just before the gubernatorial election, and Silzer's friendship and potential support are likely to have emboldened him in his break with Lindenthal. Although the enabling legislation that established the increasingly important Port Authority had been deliberately crafted to ensure its independence of political forces, Ammann would have known well that the decision makers there were not likely to be indifferent to the preferences of the governor of New Jersey.

Whatever bolstering of Ammann's confidence had resulted from his friendship with Silzer would prove well justified. Although the governor declined to campaign openly for Ammann's version of the bridge, he encouraged him to develop sketches and cost estimates, urged him to seek public endorsement on both sides of the river, and introduced him to important sources of support. It was a long and difficult period for Ammann, who was earning no income and had little taste for (and little experience in) selling himself and his work to the public. He was doubtless troubled by the unpleasant conclusion of his relationship with Lindenthal, a concern exacerbated by Lindenthal's assertion to Silzer that Ammann had abused Lindenthal's trust by profiting from knowledge gained in his employ.

But he soldiered on, dividing his time between promotional activities and a good deal of intense and productive work on his own graceful design for a lighter and less expensive bridge. At Silzer's request, Ammann submitted to him a detailed, written description of his whole concept, as well as details of the studies he had been making and estimates of construction cost. By December 1924 the governor had passed along Ammann's comprehensive report to the Port Authority, together with his own suggestion that the agency implement it and finance the bridge with tax-exempt bonds. Early in 1925, the states of New York and New Jersey passed legislation "authorizing and empowering the Port of New York Authority . . . to construct, operate and maintain a bridge across the Hudson River, from a point between 170th Street and 185th Street, Manhattan, and a point approximately opposite thereto in Fort Lee, N.J."

The Port Authority had by then begun to accumulate a workload of its own, including preliminary planning for a couple of bridges to connect Staten Island with New Jersey. It would be later in 1925 before it would get around to a formal endorsement of the Ammann proposal for a bridge at Fort Lee, but things moved more quickly after it did. Within a month or so, more legislation was enacted in both New Jersey and New York authorizing construction of three bridges, and by summer 1925, in response to Silzer's enthusiastic recommendation that Ammann be placed in charge of the work, the Port Authority had appointed him its first bridge engineer.[9]

It would be difficult to characterize the Authority's decision to place Ammann in charge of such a program as anything but a risky one that worked out very well. Although it wouldn't take long for Ammann to demonstrate that the decision was inordinately wise, there wasn't much evidence in 1925 that such an outcome was likely. His experience had certainly been broad, but only about half of it had been in genuine design work. The rest had been focused mainly on the fabrication of bridges (as in his early years with McClintic Marshall) or on the construction of bridges (as in his later years on the Hell Gate project). Of course, even the years he spent in fabrication and construction included some design work, but before the

Port Authority appointment he had never had full charge of the design of a long-span suspension bridge from beginning to end. There were certainly many other engineers around in 1925 whose qualifications for heading up the design and construction of a singularly historic bridge would have been stronger than Ammann's.

The evidence suggests that once Ammann had been given the enthusiastic endorsement of Governor Silzer, his engagement by the Port Authority needed only to be ratified by William Drinker, its chief engineer. A man whose own experience had been gained mostly in work that did not include bridge construction, Drinker would later be shunted aside when bridge and tunnel work became the agency's principal interests. For the directors of the Port Authority, the enthusiasm of the governor of New Jersey and the approval of their own chief engineer were apparently enough to convince them that Ammann was the man for the job.

One explanation of how that series of events came to play out relies on a theory that everything George Silzer knew about Othmar Ammann, when he met him in 1917, he had learned from Gustav Lindenthal. Silzer and Lindenthal had known each other during (and perhaps before) their joint service on the board of the clay mine, long before Ammann appeared. When Lindenthal proposed Ammann as the mine's new manager, his own agenda included a strong need to ensure Ammann's employment for a few years, and it's not unlikely that he described him to Silzer in a way that was compelling and favorable, perhaps even idealized. Once Ammann was hired and his performance supported everything Lindenthal had said about him, it would have been easy and natural for Silzer to become Ammann's vigorous advocate and mentor. In such a context, it's not difficult to understand Silzer's readiness to give advice to Ammann about strategies and then to attach the prestige of his own high office to an endorsement of his bridge and to a recommendation that the Port Authority place him in charge of designing and building it.

Such an interpretation suggests that Ammann's split with Lindenthal and his displacing him as the favored designer of the bridge may have originated with Lindenthal's very enthusiastic recommendation to Silzer back in 1917, adding a poignant irony to their story. Although Ammann would in a later technical review of the completed bridge acknowledge Lindenthal's "special advice on design questions," the breach between the two was deep. It never healed, and Ammann was conspicuously absent from Lindenthal's funeral eight years later.

It was during that first summer with the Port Authority that Ammann would begin to deal with the enormous task of designing and building a bridge of unprecedented span and importance. Its 3,500-foot clear span would be almost twice as long anything previously built, and some of what he would propose would rely on theory that had been only recently developed and infrequently tested. Virtually everything he did would have to endure the scrutiny and sometimes captious criticism of experts and occasionally of politicians, a prospect that had to be intimidating to a man whose work for years had been limited mostly to the execution of ideas and designs conceived by others.

His first task at the Authority's still rented first offices on lower Eighth Avenue was to recruit a staff. Unlike an established firm with the luxury of a permanent cadre of engineers that could be expanded when necessary, the Port Authority was not yet in the engineering business and had virtually no engineering staff at all. Although it had recently started work on the two bridges that would connect Staten Island with New Jersey, it had contracted out their design to a private-sector firm, and it employed only a few engineers on staff for monitoring and supervising the work.

Early along, Ammann separated his workforce into five divisions and set out to place a trusted and well-qualified engineer in charge of each. One division would explore and resolve traffic issues; one would design the bridge; one would write contracts and specifications; one would manage construction; and one would develop the approach work. Starting with those division leaders, Ammann was able to get the benefit of all their efforts and experience for recruiting subordinates. He declined to make the division separations very rigid, electing instead to allow personnel and ideas to move easily from one division to another throughout the course of the work, ensuring good communication among a variety of disciplines. Over the next few years, the size of the staff would vary in proportion to the changing needs of the project, sometimes reaching several hundred persons, including almost one hundred in the construction division alone.[10]

But Ammann hadn't come to the project to be its administrator. As soon as the process of building the staff was sufficiently advanced, he focused his attention on the design of the bridge, where his heart was. Here, he already had a solid head start. During the years between 1920 and 1923, before he left Lindenthal, he had spent a good deal of time working on his employer's design. Although that certainly wasn't the bridge he wanted to build now, it was indeed a long suspension bridge like his, with an almost identical span, and in its most recent incarnation it was being designed to carry motor vehicles on its upper deck. Certainly many of the design issues Ammann would face after 1925 were common to both schemes, and it is unlikely that many of them would be new to him. After leaving Lindenthal, he had worked almost continuously on his own design, most intensely after Silzer had begun to steer him to critical public groups (including the newly established Port Authority) for support. That period between 1923 and 1925 had been exceptionally productive for Ammann. During those years, it appears, he identified and refined the concepts that would be central to his design of the bridge.

The early tasks comprised mostly studies, some of them not much more than proofs that would provide historical justification for decisions that had already been made, but some were genuinely exploratory. In the first of those two groups were an analysis and confirmation of the basic need for the bridge, and a study of alternatives to the suspension form itself. Studies in the second group focused on fixing exact locations for the bridge's anchorages at 179th Street in New York and at Fort Lee in New Jersey, the original authorization having specified only that the New York anchorage be built somewhere between 170th Street and 185th Street. They would also refine earlier traffic studies and supplement earlier borings to provide additional, critical subsurface data.[11]

While those studies were in progress, the Port Authority began to recruit its own elite group of engineers as a consulting resource to which Ammann could turn for review of his work and for advice in matters of design and construction. Among them was William H. Burr, a Columbia University professor in his late seventies with a long and distinguished history in bridge design and construction and in the teaching of civil engineering at Rensselaer and elsewhere. Another consultant was Leon Moisseiff, a Russian-born civil engineer who, after graduating from Columbia in 1895, had been employed for a while by Professor Burr himself and had then achieved distinction in the design of many important bridges and in the development of new engineering theory. Moisseiff would have more influence on the actual design of the bridge than any of the others. General George Goethals, the former chief engineer of the Panama Canal and the first chief engineer of the Port Authority, was included in the advisory group, too, but he served only briefly before his death in 1928. A fourth consultant was Daniel Moran, a civil engineer who provided special expertise in subsurface issues of the kind that could be expected when the tower foundations and the anchorages were built. Moran was the senior partner in a civil engineering firm that would later be called Moran, Proctor, Meuser and Rutledge, which specialized in such work and focused on especially complex foundations for bridges and buildings.

Altogether, the group represented an exceptional concentration of engineering experience and knowledge, just what was needed to support the talented but still relatively young Ammann. Only a few months later, to consult on important architectural decisions, it was expanded to include Cass Gilbert, the internationally acclaimed American architect who had designed the Woolworth Building and more than a few other distinguished buildings as well.[12]

Toward the end of 1925, Ammann was finally able to give his attention to the work of the men who would actually be engaged in the day-to-day business of designing the bridge. They would be supervised by Allston Dana, Ammann's designated engineer of design, who would remain closely associated with him in bridge work for almost another twenty years. The drawings needed first would be those showing the tower foundations and the massive construction required to anchor the big bridge cables. To design them it was going to be necessary to produce reliable preliminary estimates of the loads they would have to sustain. That was a task that would mean making a first pass at designing the bridge's steel superstructure, something Ammann had been pondering for a good many months by then and about which he had some definite but sometimes unconventional ideas.

It was mainly the dead weight of the steel superstructure itself that had made Lindenthal's rejected bridge so heavy and expensive, and Ammann knew that ensuring the economic feasibility of his own bridge was going to require that he design the lightest steel frame compatible with safety. He started by addressing the matter of the live load the bridge would have to carry. Contemporaneous standards already allowed for exploiting the well-understood natural diversity of traffic by assuming that the bridge would rarely have to support more than half as many trucks and cars as its eight lanes could simultaneously accommodate. But such a reduction didn't go far enough for Ammann, who thought the reduction should be greater. He was able to convince the authorities that designing for a live load imposed by only about 20 percent of the bridge's potential total capacity was prudent (something that was shown by studies and testing to have been reasonable), and he designed the steel accordingly.

And that wasn't all. Ammann further advocated the application of deflecting theory, an increasingly well accepted concept that deflections in long suspension spans are inhibited by the dead weight of the decks themselves and by the weight of the exceptionally stiff cable and suspender systems needed to support them, obviating the deep (and heavy) trusses normally used to stiffen such spans. His bold live-load assumptions and his decision to omit heavy stiffening trusses allowed Ammann to reduce the weight of steel radically, saving what he estimated (after adding contingent tower and cable weight reductions) to be about 30 percent of construction costs and producing the elegantly slender profile that became the bridge's signature.[13] That gloriously sleek silhouette, only as deep as the deck framing itself, was for years a cherished Hudson River sight, but when a lower deck was added in 1962, stiffening trusses to connect the upper deck with the lower one pretty much concealed the original profile.[14]

Such preliminary work and further studies continued into early 1926, producing additional information about underground conditions, anchorage and tower locations, approach options, and the like. Cost estimates were refined to reflect developing design ideas, and the Authority's engineers and management were provided detailed information about present and anticipated traffic volumes and revenue potentials. Recognizing the wisdom of designing the bridge's systems to accommodate future growth but favoring the deferral of any construction that could be practically added later, the directors approved a plan designed to achieve both. It provided for the design and construction of a single deck to carry eight lanes of truck and automobile traffic, supported by foundations, towers, and suspension systems that could accommodate the addition of a second deck in the future. Those recommendations were submitted to the governors of New York and New Jersey in February 1926, and their prompt approval generated advances of a little more than $5 million from each state, to be dispensed over five years, together with an agreement that the money would be repaid, with interest, from revenues that the bridge was expected to produce. With that seed money in hand, the Port Authority was able to issue $50 million worth of bonds ($60 million had been authorized). In December 1926 the first $20 million worth of the issued bonds were sold, and just three years later, when construction of the bridge was approaching its final year, the remaining $30 million worth were sold.[15]

It has been said that Ammann considered the design of the bridge to be a relatively uncomplicated matter, that it relied simply on the logical application of fundamental laws of science, but he might have had occasion during and after 1926 to reconsider that appraisal. The purely technical complications were relatively few, but other, related issues would find their way into the design process, distracting the designers and builders and occasionally delaying their progress.[16]

Establishing the location of the New York anchorage, for example, became problematic as early as late 1925, as soon as local groups became aware that a massive (granite-faced) concrete anchorage would be built in Fort Washington Park to resist the 124,000-ton pull of the bridge's main cables. The anchorages on the other side of the river would be fully buried and secured within funnel-shaped cavities that would be drilled and blasted out of the rock and then filled with concrete, so there were no objections there; but the topography on the New York side just didn't allow for such an approach. Not surprisingly, a well-organized and articulate local group demanded that the offending New York anchorage be relocated to a more acceptable place about 600 feet farther east, a change that would have meant carrying the bridge's structure and all four of its giant suspension cables across Riverside Drive and beyond, giving the whole design an unwanted asymmetry. In addition, such a move would have meant reworking the approach layouts, which in turn would have delayed construction by about a year and added what Ammann estimated would be more than $7 million to the cost. The controversy went all the way to the desk of New York's governor before being resolved, two years after it had surfaced, by an agreement to allow the anchorage to be built where originally planned in exchange for a like piece of land elsewhere and enough money to landscape and develop it as a park.[17]

Meanwhile, work on the bridge's design was proceeding apace, punctuated by occasional sniping from political critics who disagreed with decisions that were being made and from a few engineers who continued to regard Ammann as insufficiently experienced for the responsibility he had been given. For the most part, though, he was proving himself to almost everyone. In a 1926 reorganization of the Port Authority, William Drinker, who until then had been chief engineer, was designated "chief consulting engineer" and separated from the bridge group, a reassignment that clarified Ammann's absolute authority for all bridge work. A year later, Ammann was given the title of "chief engineer."

Just about everything connected with a bridge of such enormous size was bound to be big, and the concrete foundations for the towers were no exception. These were the piers on which the big tower legs would stand, approximately 600 feet from the anchorages. The two piers on the Jersey side, where the first tower foundations were built, were almost 100 feet square and would be founded on rock about 80 feet below the surface of the water. The open cofferdams needed to allow their construction under comparatively dry conditions would be among the largest ever built.

The contract for the difficult and dangerous job of building those huge foundations, the first of twenty construction contracts to be let for the bridge, was awarded to the Silas Mason Company of New York in the early spring of 1927, and by May the work was under way. The Mason company, which would later go on with partners to build the Grand Coulee Dam and the Lincoln Tunnels, had won the job with a bid that was well under most of the others and less than half the highest. By December, Mason's crews had almost finished driving the sheet piling for the first cofferdam when an unanticipated rock fault caused a catastrophic failure in several of the sheet piles, flooding the cofferdam in less than five minutes and drowning three men. It was a sobering tragedy, especially at the very beginning of the work, and there would be nine other deaths from work-related accidents before the bridge was finished. But the inevitability of tragedy in such hazardous work seemed to be well understood, and construction was resumed quickly. Within about two years foundations for the Jersey tower and rock excavation for the Jersey anchorages would be finished.[18]

By spring 1928 contracts representing more than half the estimated direct cost of bridge construction had been awarded. In addition to those for the Jersey foundations, these contracts were for the New York tower foundations and anchorages, for both the steel towers, for the bridge deck, and for the largest project of all, furnishing and erecting the cables and suspension systems. It had all been coming together nicely, as far as the budget was concerned, but not without an occasional setback. A few problems, not all of them technical, had continued to roil an otherwise smooth process, and resolving them in time to put all those contracts in place by the early months of 1928 hadn't been easy.[19]

Deciding just how the two 604-foot-high towers would look when the bridge was finished was one of the problems. Ammann would later write that no aspect of the design had elicited as much comment "both favorable and unfavorable" as the towers, and that the "natural beauty of a graceful, suspended structure" could be enhanced or destroyed by a decision as to whether to encase the steel framing of the towers in masonry or to leave it exposed. Early in the design process, there appeared to be a preference for the monumental aesthetic of a composite concrete and steel tower structure, in which the two materials would share the loads. But in later studies preference shifted toward an independent steel frame with full capability for supporting all the tower's loads, without the benefit of any concrete. Such a self-sustaining steel structure was substituted for the composite one, and the detailing of the tower was done to allow for the later addition of a masonry enclosure, if and when desired.[20] After that change had been made, there was a good deal of discussion about whether the enclosure should ever be added and about whether it should be built of concrete or stone or both. And the question of just what stone it should be further complicated matters.

Cass Gilbert, the consulting architect for the bridge, vigorously advocated a stone enclosure, and he identified just the stone he wanted to use. Leon Moisseiff, the principal structural consultant on the design of the towers, favored leaving the steel exposed. Ammann himself acknowledged a preference for "encasement with an architectural treatment, such as that developed by . . . Cass Gilbert," and he dismissed philosophic objections that encasement would conceal the proper expression of the structure's real function. Having made that clear, he later wrote that he was pleased with the appearance of the unencased structure, "which owes its good appearance largely to its sturdy proportions and the well balanced distribution of steel in the columns and bracing." And it was just as well that he was happy with the exposed steel, because the towers were never encased. The economic pressures of the Depression were starting to be felt during the last days of construction, and there was little enthusiasm for spending any more money than absolutely necessary.[21]

That wasn't the only design-related controversy that had to be resolved before all the main bridge contracts could be awarded. A difference of opinion about the design and construction of the huge cables that would carry the bridge deck had surfaced almost as soon as the first sketches of Ammann's concept emerged in 1925. Some well-informed opinion (including Ammann's) held that the main bridge cables should be formed of linked eyebars, while others favored the use of parallel wire cables. Eyebars are thick, flat, steel bars with holes in each end to allow them to be linked with adjacent, identical bars to form chains. Such chains already supported many suspension bridges, including some historically important ones. Lindenthal, in fact, had used eyebar chains for several of his bridges, and when he was bridge commissioner in New York, he had made a vigorous (but unsuccessful) attempt to redesign the wire cables of the Manhattan Bridge as eyebar chains. Even the design for his proposed Fifty-seventh Street bridge (which Lindenthal was still advocating) was based on using eyebar chains, so it was clearly not an obsolete practice. But starting about 1816, cables made of multiple strands of parallel wire instead of eyebars had begun to gain acceptance, especially in Europe. John Roebling, an American engineer who manufactured wire in his own plant, had begun using it in his bridges as early as 1844, and once he specified it to carry the long deck of his historic Brooklyn Bridge around 1876, it increasingly became the preferred material for most long-span suspension bridges.[22]

Normally, a debate about whether wire cables or eyebar chains would be used for bridge suspension wouldn't stir much passion. The arguments would probably be mostly technical, occasionally punctuated by aesthetic concerns. But in the case of the bridge that would connect 179th Street with Fort Lee, commercial and political factors enlivened and lengthened the debate. The Roebling Company (by then managed by descendants of the founder) was a major manufacturer of wire, not eyebars, and its plants were in New Jersey, where it employed a great many workers. The idea that a design decision might prevent the company from competing for a very large contract for what was to a great extent a New Jersey enterprise was understandably objectionable to Roebling's management. In 1927 the company's president brought the matter to the attention of A. Harry Moore, who had succeeded George Silzer as governor of New Jersey. Moore was no friend of the Port Authority, resentful of its brash independence, and he was pleased to have a pretext for bringing it to heel. He quickly persuaded the legislature to pass a bill that required all Authority contracts to be reviewed and approved by New Jersey's State House Commission, a group historically linked to the distribution of political patronage. Governor Al Smith of New York, together with a respectable number of like-thinking people in both states, immediately recognized the hazard of allowing politicians of either state to exercise control over the professional work of the Port Authority, and they were able to induce the repeal of Moore's bill before any serious harm was done.[23]

Meanwhile, Ammann had brought his own imaginative perspective to the dispute, combining sound engineering with some good commerce. He revised the drawings and rewrote the bid invitations to allow competing contractors to submit proposals based on either the eyebar design or the parallel wire design (and offering a few other variations, too), opening up the competition and eliminating any basis for dispute. Five bidders responded with proposals that offered the Authority fourteen different combinations to choose from. When the proposals had been opened and read aloud, a combination of the Roebling Company's approximately $12.2 million bid to do the suspension work using wire cables and the McClintic Marshall Company's approximately $11 million bid to do the rest of the steel work was recognized as the cheapest, and within a month those two firms had been awarded contracts. There would be no eyebars.[24]

The opening of those major bids was a watershed in the early history of the bridge, confirming to any lingering doubters that Othmar Ammann had indeed been an exceptionally good choice. Not only had he produced a universally popular design, he had also dealt with a potentially obstructive dispute in a graceful and effective way and had reduced the cost of construction in the process. There had been some evidence that eyebars might in fact prove to be the less expensive option, but Ammann was satisfied that either system would do the job well, and he had elected to secure the best price by relying on Roebling's fierce and widely known determination to get the job to drive the price down. And that wasn't all. The signing of these big contracts with Roebling and McClintic Marshall raised the total value of all contracts signed to date to about 80 percent of the approximately $32 million that had been estimated to be the base cost of building the bridge itself (before the cost of approaches, land acquisition, or interest). With so much of the most volatile component of cost within budget, it was virtually certain that the job would be completed without any need for additional funds, a prospect that clearly enhanced the reputations of the Port Authority and its principal engineer.

Even with both those big superstructure contracts signed, there was still plenty of work to be done, both on-site and off-site, before any steel would be erected or any cable would be spun. It would be more than six months before work on the second of the two piers for the Jersey tower would be finished, a delay at least partially attributable to bad weather, and at the end of 1927 there was still no contract at all for building the foundations for the New York tower. But the critical path of the work was unlikely to be severely deflected by either of those scheduling issues, because both the steel contractor and the wire cable contractor would need every one of those intervening months to assemble and prepare their materials.

McClintic Marshall would be the first to perform, but before its crews could start erecting the towers, most of about 40,000 tons of structural steel would have to be detailed, fabricated, and shipped to the job. Roebling would be right on McClintic Marshall's heels with its cables, once the towers were in place, but not before a substantial fraction of its own 30,000 tons of wire had been manufactured, tested, and delivered to the site.

Before the 1927-1928 winter ended, the Port Authority had at last (and not a moment too soon) signed a contract with the Arthur McMullen Company of New York to build the foundations for the New York tower and to build the controversial east anchorage in Fort Washington Park. By spring, when the Jersey piers were finished and work on the huge but shallower and easier New York piers was beginning to show good progress, signs abounded that steel work would soon start. The Jersey side of the river was coming alive with barge-mounted derricks and car floats towed up the Hudson from terminals in New York and New Jersey, loaded with steel fabricated in Pennsylvania and shipped east by rail. By the summer of 1928 the whole area had taken on the appearance of the genuine spectacle it would soon become, as ironworkers began erecting their huge fabricated column assemblies, some as heavy as 100 tons apiece, on the concrete piers. Over the months that followed, traveling cranes weighing more than 300 tons would claw their way up the sides of the very steel they had just erected, setting big steel assemblies in place, one after another, with riveting gangs following impatiently.

It wasn't long before the Port Authority found it necessary to build viewing areas along the shoreline to accommodate the crush of spectators, and the press began following the progress on an almost daily basis. There were very few significant delays. At one point, the erectors got a bit ahead of the fabricators, and it was necessary to slow the pace of fieldwork for a week or so to give the plant a chance to catch up. And when the ironworkers were within only a few lifts of topping out the towers, there was another brief stoppage to allow precise measuring of what had been erected, just in case corrective adjustments in the fabrication of the remaining steel needed to be made. The field measurements were said to reveal a maximum deviation from design dimensions of only three-sixteenths of an inch, and by June 1929 the towers were ready for cable, and the steel contractor left the job, set to return as soon as Roebling's crews finished.[25]

Few of the bridge's features dramatize its gigantic scale as effectively as the widely quoted statistics that it took 107,000 miles of 0.196-inch-diameter wire to produce the four gigantic cables that carry the bridge deck itself, and that the four cables together weigh about 30,000 tons. Once the approximately 600-foot-high towers were in place, Roebling's crews began unreeling and spinning the cable wires, spiriting them at high speed back and forth across the bridge, from one anchorage to another, over the massive steel saddles that had been integrated into the tops of the towers, until they had all been spun into the four enormous catenaries that would carry the deck. Compacted hydraulically and wrapped with wire, the big cables had been made ready for long lives above the river. Two of them were positioned 9 feet apart to support the north edge of the deck and the other two, centered 106 feet away from the centerline between the first two and also positioned 9 feet apart, would support the south edge.

The difficult and dangerous work of hanging the deck from those huge cables would now be left to the crews of the structural steel contractor. In preparation for this critical task, Roebling's cable crews had hoisted into position a system of multiple, 3-inch-diameter, wire rope suspenders and draped them over each of the supporting cables every 60 feet, from anchorage to anchorage. The long suspender ropes dangled like windblown garments on a clothesline, hundreds of feet above the Hudson, while the structural steel contractor began preparations below.

There was something spectacular about almost everything that was done to erect the bridge, but the Roebling Company was clearly the star of the show. It had built the Brooklyn Bridge fifty years earlier and had played an important role in developing almost every one of the concepts and techniques that were essential to building the new, very long spans that followed. Roebling's crews, averaging more than 300 men on this job, had first spent three months building and erecting a pair of suspended footbridges as working platforms, and then they had worked from those platforms to spin the cables for the bridge itself in about ten months. That was about half the time it had taken to spin the Brooklyn Bridge's cables, which had used only about one-fifth as much wire.[26]

During the first week of August 1930, fireworks announced the completion of the cable work. After a few hours of hard-earned celebration, Roebling's crews packed up and left, to be replaced by McClintic Marshall's ironworkers and its fleet of barges and derricks. Undaunted by the onset of winter, the McClintic crews started early in November, and by January 1931 they had erected the 17,000-ton deck 200 feet above the Hudson and were on their way back to Pottstown.

With the deck in place, the end of the job was finally in sight. As soon as the weather began to warm up in the spring of 1931, the road surface was paved: six lanes for current traffic, with a swath left unpaved in the middle for two more lanes to be added when circumstances required. There were still almost a dozen small contracts to be completed, but only a few were essential for the opening of the bridge. The Port Authority's management, exulting in its success, began talking off the record about advancing the opening date, which had until then been scheduled for the middle of 1932.

As with all great projects, there were occasional, important problems that didn't require great expertise for solution. As opening day drew closer, one of the looming ones was the selection of a name for the new bridge. There was no shortage of suggestions, including the Verrazano, the Columbus, the Fort Lee, and the Palisades. There was even some enthusiasm for calling it the Hudson River Bridge, a name that had already caught on a little. But the strongest support early in 1931 appeared to be for naming it the George Washington Memorial Bridge. In January the Port Authority responded to that apparent preference by ratifying the name, but it didn't stick. Most people agreed with the idea of honoring the country's first president, but some argued that there were already bridges and other structures in the

FIGURE 10 The George Washington Bridge, in a photograph taken during the early 1930s, long before a second deck was added. Photograph courtesy of the Port Authority of New York and New Jersey.

region with similar names. In April the directors of the Port Authority resolved to drop the word "Memorial" and settle on simply the George Washington Bridge.[27]

By the twenty-third of October 1931, the bridge was substantially complete and ready to open, about eight months earlier than its five-year target and about $2 million under its $60 million budget. Opening ceremonies came the next day, when a crowd estimated at 30,000 gathered to hear the governors of New York and New Jersey extol the virtues of George Washington, acknowledge the splendor of the new bridge, and praise the Port Authority commissioners and a few politicians, without paying much attention at all to the engineers and workmen who had designed and built the bridge. But the crowd seemed to know better. The loudest and most prolonged ovations of all were said to have been given when Othmar Ammann and the aging Gustav Lindenthal, eighty-one by then, were introduced.[28]

In 1946, soon after the end of World War II, the unpaved middle of the bridge deck was filled in, increasing the original six lanes to eight. Later, in 1955, when a joint commission of the Port Authority and the Triborough Bridge and Tunnel Authority recognized that there was a serious need for more capacity, a lower deck providing six additional lanes was added. Completed in 1962, it relieved a monumental impediment to the flow of traffic between New York and New Jersey, but it also obscured the slender profile of Ammann's original bridge, disappointing many who had revered it.

And although the lower deck cost $183 million, more than three times the cost of the original bridge, it demonstrated its value within only a few years. Ammann had originally predicted that an average of three million cars would cross the bridge annually in the early years but that in twenty years the figure would grow to twelve million, with like increases to follow. In 2007, seventy-six years after the bridge was opened, more than one hundred million vehicles crossed it.[29]

CHAPTER 9

The Mid-Hudson Bridge

It's no accident that the decade of the Roaring Twenties was the heyday of bridge and tunnel building in the lower Hudson Valley. During those ebullient years between the end of World War I and the beginning of the Great Depression, when the Holland Tunnel, the Bear Mountain Bridge, and the George Washington Bridge were built, the country was in an extraordinarily expansive and optimistic mood. New social and political ideas abounded, new concepts were emerging in architecture, engineering, painting, literature, and music, and a feeling spread across America that the good times were likely to last forever. There was more enthusiasm for grand projects than there had been in most earlier years, and it was a good deal easier to obtain support for them than it would be just a few years later. By most accounts it was a wonderful time for proposing a new bridge.

Bridge construction was especially well supported in New York City during the twenties, but that wasn't the only place. Seventy-five miles up the Hudson, at about the time work on the Bear Mountain was winding down and about when Lindenthal and Ammann were beginning to disagree about the design of the George Washington, plans for a second bridge at Poughkeepsie were taking shape.

It had been only about thirty-four years since completion of Poughkeepsie's historic railroad bridge, and the reason for building a second bridge within less than a mile of the first one was seen by some as frivolous. Its proponents just wanted to be able to walk across the bridge, instead of riding on a train, and they were campaigning for alterations that would accommodate a pedestrian walkway. The existing bridge already had a couple of walkways, but they had been designed to provide inspection and maintenance access for railroad employees and were considered too narrow for any other use.

Ten years earlier, the mayor of Poughkeepsie had raised the possibility of improving those limited (and sometimes dangerous) walkways. He had been able to attract some enthusiasm for his idea from Monsignor Joseph Sheahan, the activist pastor of Saint Peter's Roman Catholic Church in Poughkeepsie, and additional support from his own influential brother, a member of the powerful Public Service Commission of New York State. But the idea didn't get much traction. Although the charter for the original bridge was interpreted by some as obligating the railroad that had acquired the bridge to provide pedestrian access, a legal review turned up a revision to the charter that had pretty much relieved the company of any such responsibility.

But the mayor had planted a seed, and Monsignor Sheahan took on the job of nourishing it. The pastor became convinced that what was really needed wasn't just pedestrian access, but vehicular access, too, and over the years that followed he sponsored and encouraged a series of petitions to improve or expand the old railroad bridge. None of them succeeded. By 1921 Monsignor Sheahan and his following had found their way to Gustav Lindenthal, who was at that time gearing up for what proved to be his final, failed effort to secure the commission to design the George Washington Bridge. What they wanted from Lindenthal was an expert opinion to support their idea that vehicular and pedestrian access could be provided by reinforcing, altering, and adding to the old bridge. But they were to be disappointed. Lindenthal, not a bit reluctant to disparage the original design, inspected the old bridge and told them that the cost of such an improvement would be so high that no responsible engineer could justify it. It became evident that if the citizens of Poughkeepsie were serious about vehicular and pedestrian access for crossing the Hudson, they were going to have to build a new bridge.

Apparently, they were serious. The evolution of their objective from simple pedestrian access to vehicular access and then their leap to building a new bridge had broadened and intensified the appeal of their campaign. The idea caught on, and in little time the local newspapers and the growing local chapter of Kiwanis, together with some of the town's most influential citizens, had lined up behind Monsignor Sheahan to advocate for a new bridge.[1]

In 1922 they established the Hudson Valley Bridge Association to mount and manage a campaign. There would be no Harriman family to pay for this bridge, and there would be no venturesome investors or expanding railroads to underwrite it, either. The bridge advocates, with insufficient assets of their own, would have to look to the state to build their bridge, and they focused on establishing a case for it. They started with an engineer in New York named Theodore Pratt, secretary of the fledgling New York State Motor Truck Association, an organization that saw a vehicular bridge at Poughkeepsie as a means to capture some of the railroads' freight business. Pratt did a good job, traveling the Hudson Valley to probe patterns of commerce and traffic and writing a report that quantified the huge and growing volume of farm products that would benefit from shorter routes to their markets. He described the new facilities New York City was building to receive and distribute such products and dramatized the unreliability and inadequacy of the existing ferries. Little of what he wrote about the historical role of bridges in the growth and economic well-being of river towns was lost on the movers and shakers on either side of the Hudson.[2]

Once the case for building a bridge had been made, the association needed to demonstrate to potential patrons in the state capital that such a bridge was feasible. For that purpose, they engaged George W. Goethals and Company, Consulting Engineers, and it was a strategy that proved to be inspired. In 1922 General Goethals, the retired army officer who was principal of the company, was known to just about every literate adult in the United States. Fifteen years earlier, when he was a relatively young member of the U.S. Army's prestigious General Staff, he had been sent to Panama by President Theodore Roosevelt with instructions to build the canal that the French had abandoned, and he'd been given all the money and authority he'd need to do it. Seven years later, the Panama Canal completed, he served as first governor of the Canal Zone. When that term ended in 1916, he retired from the army as a major general. He was a national hero by the time the bridge association engaged his firm, and the committeemen are likely to have reasoned that it would be difficult for any state official to reject a recommendation from him about a major engineering project. (They were probably unaware that Clifford Holland had managed to do just that without dire consequences only a few years earlier, when he rejected the Goethals-O'Rourke plan for building the Holland Tunnel.)

Goethals was born in 1858 in Brooklyn of working-class parents who had emigrated ten years earlier from the Flemish-speaking Belgian city of Ghent, where they had been able to trace their history through generations of crusading knights and other noblemen, all the way back to ancient Rome.

Young George proved to be a superior student and studied at City College in New York for a few years, but withdrew before graduating in order to accept an appointment to the U.S. Military Academy at West Point. There he was successful and popular, graduating second in his class in 1880 and becoming one of only two graduates to be assigned to the Corps of Engineers. He spent approximately the next twenty-six years of his life executing a variety of engineering assignments that focused mainly on water and sanitary projects. When Theodore Roosevelt dispatched him to Panama in 1907, it was on the recommendation of William Howard Taft, his secretary of war, who had come to know Goethals from his work on the army's General Staff.

Goethals's retirement years were less successful. Accustomed to managing with absolute authority, he served with some difficulty on a number of commissions and boards, and ended up resigning from several of them. By 1922, when the bridge association engaged him for the work at Poughkeepsie, he had established his own consulting firm in New York.

Goethals explored a range of options for a new bridge at Poughkeepsie and recommended a location less than a mile south of the original one. Apparently daunted by the prospect of designing a suspension bridge, he recommended and prepared a well-developed preliminary design for a through-type, riveted cantilever truss bridge consisting of five main spans of 540 feet each, with four piers in the water, aligned approximately with the piers of the existing bridge. His sketches showed 140 feet of clearance between mean high water and a concrete roadway at the level of the lower chords of the trusses, and he proposed a caisson system for building the concrete piers to support the trusses. He estimated a cost of about $5 million for the construction.[3]

Meanwhile, the association had lined up some formidable support in the state legislature. Although the best they could do in the 1922 session was to secure some kind words and encouraging predictions, they would be able to obtain passage of the required enabling legislation in the 1923 session, together with an appropriation that would see them through the engineering work.

The decision to engage Goethals worked out very well, and almost everyone important to the project appeared to like his report. Chances are that the committee that selected him had been unconcerned or unaware that the only bridge George Goethals had ever designed was an approximately 120-foot-long wooden structure built by soldiers in the State of Washington about forty years earlier, when he was only a couple of years out of West Point.[4]

Fortunately for the bridge's advocates, Frederick Stuart Greene, who as chief engineer of the New York State Department of Public Works was the ranking engineer in the state's technical hierarchy, was among those who found the Goethals report satisfactory. Greene was responsible for evaluating the bridge's feasibility for the state and would be the man in charge of getting it built. An 1890 graduate of the Virginia Military Institute, Greene had himself retired from the U.S. Army as a colonel before entering state service. A fairly complex personality, he was by 1923 well along in an eminently successful second career as a writer of horror fiction, a genre in which he already had several dozen titles to his credit. His work would still have an audience eighty-five years later, when Cemetery Dance Publications of Forest Hill, Maryland, included a story of his in its two-volume anthology of the best horror fiction of the twentieth century.[5]

Although Greene was an admirer of Goethals and was satisfied with his feasibility study, he wasn't inclined to engage the general to design the bridge at Poughkeepsie. He turned instead to the nationally prominent bridge designer Ralph Modjeski. Sixty-two years old when he and Greene sat down in 1923 to talk about Poughkeepsie, Modjeski was an engineer of outstanding reputation and uncommonly broad experience. Born in Poland and educated in France, he had come to America in 1885. Within a few years of his arrival he had worked on at least six major spans, including the historic Frisco Bridge across the Mississippi at Memphis, said to be the longest cantilever bridge in the country when it opened in 1892. During the years that followed, the pace of his work accelerated, and by 1912 Modjeski had completed more than a dozen other big bridges, including a double-track railway bridge at Thebes, Illinois, the McKinley Bridge at St. Louis, and the spectacular Broadway Bridge in Portland, Oregon, which at the time was said to be the largest bascule type bridge in the world. By the time of his meeting with Greene, Modjeski had designed bridges in Connecticut, Illinois, Iowa, Missouri, North Dakota, Ohio, Oregon, Pennsylvania, and Tennessee, and he had served as a bridge consultant in Alaska and Canada. Few engineers in the country could count more major projects to their credit.

Like Greene, Modjeski found personal fulfillment beyond his engineering work. Trained as a classical pianist before he turned to engineering in his late adolescence, he would continue to play privately and for small audiences during most of his life. Extroverted and gregarious, he integrated performing into his complex persona, probably a natural consequence of being the son of one of Poland's most acclaimed actresses, Helena Modjeska, whose surname—in the Polish tradition—was spelled with the feminine ending.

Poughkeepsie wasn't new to Modjeski when he arrived there in 1923. Sixteen years earlier, as a consultant to the New York, New Haven and Hartford Railroad, which had by then acquired ownership of the Poughkeepsie Railroad Bridge, he had designed and supervised a program of reinforcement for that prematurely aging structure. That was probably when he first became aware of some of the problems that had faced John O'Rourke and his staff when they struggled with construction of the bridge's piers in the 1880s.

It's likely that O'Rourke's detailed and cautionary accounts of the great depths, erratic subsurface topography, and sometimes treacherous currents of the Hudson influenced Modjeski's decision to invite Daniel E. Moran to form a partnership with him for designing the second bridge at Poughkeepsie. Modjeski's own history suggests that he was never reluctant to enter into such collaborations whenever he saw the need to supplement his own considerable skills or experience, and Moran, an expert in bridge foundations, offered just the special capability that Modjeski wanted at Poughkeepsie.[6]

Only a few years younger than Modjeski, Moran had been born in New Jersey but had lived most of his childhood years in Brooklyn. After a few years at the Brooklyn Polytechnic Institute, he entered the civil engineering program at Columbia University, where he played football, edited the college newspaper and yearbook, was a fraternity member, and still managed (somehow) to graduate in 1884. For half a dozen years after that, he worked his way around the country doing entry-level civil engineering jobs. By about 1890 he had become seriously interested in foundation construction and took a job with the contractor who was building the McCombs Dam Bridge across New York's Harlem River. It was on that job that Modjeski had invented an air lock that in one form or another would become standard wherever pneumatic caisson construction was required for subaqueous foundation work.

After the McCombs Dam Bridge job, Moran made foundation design the focus of his career, and in 1901 he established, with a couple of colleagues, a contracting firm that would specialize in (and become extremely successful at) the construction of deep and difficult foundations for bridges and buildings. Nine years later, he separated from that company to establish, with Princeton-educated civil engineer Ralph Proctor, a consulting firm specializing in the design and supervision of such work. Later, as the Moran and Proctor name changed periodically to reflect changes in the company's principals, the firm that Moran had established in 1910 would become and remain one of the world's leading authorities in its specialty.[7]

Modjeski and Moran were a couple of very busy engineers in 1923. Each was juggling a variety of commissions that might have overtaxed lesser practitioners. Modjeski was already serving as chief engineer of the Delaware River Joint Commission and was about to assume responsibility as chief engineer for the Delaware River Bridge itself. Moran was awash in consulting work on foundations for some of New York's burgeoning community of high-rise builders and would soon provide consulting services to the Port Authority for the George Washington Bridge. But the two of them were industrious, skilled, well supported by staff, and apparently eager to take on the task of designing what was by then being called the Mid-Hudson Bridge at Poughkeepsie. Later in 1923 they signed a design contract with the state.

Very busy or not, by early 1924 Modjeski and Moran were able to give Greene and the others a first good look at what they were proposing. It was a slender and graceful suspension bridge with two piers in the river, each 750 feet from shore, 1,500 feet from one another, and designed to support steel towers that would rise 280 feet above mean high water. Its deck was to be about 30 feet wide, from curb to curb, with a 54-inch sidewalk outboard of each curb, 135 feet above the water. The state legislature approved the preliminary documents and added to an earlier appropriation, but declined to provide all the required money because it (correctly) anticipated that the governor would soon decide in favor of a state bond issue for funding the bridge and other permanent improvements.[8]

Authority to proceed with construction was still dependent on the War Department, which had final jurisdiction when it came to bridges across navigable waterways. Although the department hierarchy liked the bridge, it couldn't approve construction until it had considered a local complaint that the position of the easterly pier would interfere with shipping into and out of the Poughkeepsie dock area. The commanding general of the Army Corps of Engineers chose a direct if informal approach to the matter. Without informing any of the complaining ship captains, he traveled to

Poughkeepsie himself and personally observed shipping patterns there for a few days, to see whether the pier would really be an intolerable obstruction. After determining that the complaint was not justified, he signed off on the bridge. By later in 1924 all the approvals needed from public authorities had been issued, and the way was clear for proceeding with the task of converting conceptual and preliminary drawings to detailed documents from which bids could be prepared and the bridge could be built.[9]

During 1925 a good deal more work was done in the office and the drafting room than in the field. Designs for work on the east and west banks were completed first, and by late spring documents for the abutments and anchorages and for the viaduct that would cross the New York Central's tracks were sufficiently advanced to allow for competitive bidding. By fall, fieldwork on those relatively small but difficult early contracts was underway, with crews on both sides of the river gearing up to drill a hundred feet into rock to secure the anchorages. Then, on the ninth of October, in an unrelenting rainstorm that failed to dissuade a large, cheering crowd from taking advantage of the holiday decreed by the mayor, Governor Alfred E. Smith laid the bridge's cornerstone.[10]

Another eighteen months would elapse before anyone outside the project's small circle of informed insiders would see clear evidence out in the river that a new bridge was really in the making. But the pace of the preparations was picking up, and plenty of progress was being made ashore. By the end of 1925 the state legislature would approve a budget of $5 million and ratify the governor's preference for funding it with state bonds, and only a few months later, New York's Department of Public Works (later to be called the Department of Transportation) would invite bids for the main contracts. This relatively modest vehicular bridge at Poughkeepsie wasn't really big enough to generate the competition that larger and potentially more profitable bridge jobs were attracting, but the bids proved to be close enough to budgeted amounts to allow for awarding the major contracts. By June 1926 the Blakeslee Rollins Construction Company of Boston had won the contract for foundations, and a contract for structural steel work, including the towers, the cable system, and framing for the bridge deck itself, had been awarded to the American Bridge Company of Ambridge, Pennsylvania. The schedule anticipated that off-site preparatory work by the contractors would take almost a year, that substructure and superstructure work would take about another two years after that, and that the bridge would open for traffic around the middle of 1929.

Blakeslee Rollins's main job would be to build a couple of massive concrete caissons, sink them to the bottom of the river, about 60 feet below the water's surface, and then drive them about 75 feet into the silty clays and boulders under the river's bed until solid material prevented their being driven any farther. Each caisson would by then have been built up in place to the height of a modern eight-story office building, weigh about 40,000 tons, and have a footprint almost three times the size of a standard tennis court. Once each caisson reached its final position, Blakeslee Rollins would fill the big cells within it with concrete and add a couple of 66-foot-high, stone-faced concrete piers, one above each caisson, to support the steel towers that would follow.

Caisson methodology was still relatively crude but well tested and effective, integrating the difficult excavation needed for such big foundation work into the actual construction. On the Mid-Hudson job, work on the base unit for the west caisson was started early in the 1926-1927 winter almost a hundred miles from Poughkeepsie, at the yards of the Staten Island Shipbuilding Company (later a division of Bethlehem Steel Company). The base would look like a gigantic concrete egg crate, 136 feet long, 60 feet wide, and 20 feet high, with multiple 2-foot-thick concrete dividing walls within and a 3-foot-thick concrete perimeter wall. A concrete and steel cutting edge that would form the lower edge of the perimeter wall would make it easier to drive the huge base section and its towering superstructure down into and through the mucky soils under the river.

The enormous, 20-foot-high base sections for the west and east caissons were the Staten Island Shipbuilding Company's only contribution to the Mid-Hudson job. The hard work of building up those 2- and 3-foot-thick walls to heights that would finally exceed 100 feet would be done out in the river, at Poughkeepsie, by Blakeslee Rollins. Although most of the cells in the egg-crate structure would be kept open to allow barge-mounted equipment to thread clamshell buckets through them, to dredge out the material below the caisson, some cells had to be closed off from the beginning to give the units the buoyancy needed to float them up the Hudson to the bridge site.

The first floating caisson base was towed from Staten Island to the west pier site in March 1927, and the work of building up its concrete walls was started there as soon as the base's position in the water could be stabilized by anchors. Raising the walls was done in sixteen-foot increments, increasing the weight of the growing caisson by about 400 tons for every foot of added height. Periodically, as the big structure's increasing weight forced it down into the river bottom, clamshell buckets would be threaded through its cells to muck out underlying material in its path. Gradually, as the caisson grew taller and heavier, the weight of the added concrete (and the continuing dredging) would drive the whole gigantic assembly down to a level about 115 feet below the water's surface. By June, that had been accomplished, and the caisson had come to rest on ledge rock and been filled to the top with concrete.

Things didn't go nearly as well for the east caisson. Its first section was towed up the river to Poughkeepsie in the spring of 1927, and work on building up its concrete walls was started there without delay. But about three months later, when its cutting edge was only about 11 feet into the mud, disaster struck. During the night shift, on the twenty-seventh of July, one edge of the east caisson sank about 29 feet, while the opposite edge rose about 11 feet. The huge structure came to rest in about a minute, listing severely, at an angle of about 43 degrees from the vertical.

Emergency measures were taken immediately to prevent further tilting of the still incomplete caisson, whose weight had by then been increased to about 19,000 tons. Anchored vessels were positioned along its easterly edge, and 150-ton anchors were sunk and secured to its westerly face by steel cables. Over the weeks and months that followed, half a dozen strategies failed to budge the massive structure, but by May there were signs that success might be at hand after all. Between multiple block-and-tackle systems rigged through gallows frames and a variety of excavating systems designed to remove as carefully as possible the material that was resisting the righting effort, something good was finally happening. An engineer monitoring the work with a transit from a shore position about 300 yards from the pier site spotted and signaled the first few inches of movement.

After that, the process of restoring the east caisson to its proper position, both vertically and horizontally, was steady but agonizingly slow. A 12-foot-diameter painted display was set up on the beach and updated daily to keep the men who were working out in the river informed about their progress. By November 1928 the east caisson had been righted. After a winter shutdown, work on building up its concrete walls was resumed, and by the end of March 1929 its cutting edge had reached bottom, 135 feet below the surface of the water. More than a year had been lost.[11]

The Blakeslee Rollins crew was doubtless more than ready to head back to Boston after their ordeal in Poughkeepsie, but they still had more work to do. As required by Moran's design, they had troweled off the approximately 8,000-square-foot top surfaces of the caissons at a level that was still about 30 feet below the surface of the water, and that meant that the lower half of the approximately 60-foot-high, stone-faced concrete piers for which they were still responsible would have to be built under water.[12] Fortunately, the timely completion of the west caisson had allowed them to get its pier built, but installing a watertight cofferdam on the top surface of the ill-fated east caisson and then building a pier inside it was something that would add just one more unwelcome delay to an already beleaguered schedule.

By the time the east pier was finished, the steel contractor, still waiting on the sidelines, was more than ready to get on with its part of the job. The Bridge Company, as American Bridge Company was often called in the industry it dominated, had plenty of reason for being out of patience. Almost two years earlier, when its managers had first heard about the caisson problem, they had told the state that they'd like to defer fabrication of their steel until the foundation contractor had solved the problem. But the state, uncertain about how long the delay would last, had insisted that work on preparing and shipping the steel proceed as originally scheduled. As a result, the Bridge Company fabricated all 10,000 tons of it and delivered it to a storage yard about five miles downriver from Poughkeepsie, where most of it would languish for more than a year. By spring 1929, when barging the steel up the river to the bridge site could finally begin, the prime coat of paint that had been applied at the shop had begun to peel, and everything had to be repainted. Altogether, it was an inauspicious start.

But things improved quickly. Erection of the 4,000 tons of steel that went into the two towers, most of it placed from barges in the river, was started in late April and completed before the end of August, a formidable achievement that made it possible for the footbridges for erecting the suspension cables to be put in place by early October and for the spinning of the cables themselves to be well advanced before the start of the 1929–1930 winter.[13]

There would be two 17-inch-diameter main bridge cables on the Mid-Hudson, each about 3,800 feet long from anchorage to anchorage, and each comprising almost 1,900 miles of galvanized wire a little over three-sixteenths of an inch in diameter. It had all been furnished by the American Steel and Wire Company, a sister company of American Bridge in the huge U.S. Steel conglomerate, and installed by the Bridge Company. But even when the cable spinning was substantially completed in the spring of 1930, almost 6,000 tons of deck framing and stiffening trusses remained to be erected.[14]

Opening ceremonies had been optimistically scheduled for 25 August 1930, and the race to finish in time would be a close one, with the Bridge Company successfully extending itself to make up for time lost by others. Well into July, concrete for the bridge deck was still being poured (in alternating sections, mainly to maintain balanced loads on the cables and towers), and the sound of rivet guns could still be heard.[15]

The bridge was finished by the scheduled opening date (except for a few small details). Twenty-five thousand spectators crowded into and around the reviewing stands near the east entrance to the bridge to hear former governor Alfred E. Smith (who had arrived by yacht) and his successor, Franklin D. Roosevelt (who had been driven to Poughkeepsie from his home in nearby Hyde Park), praise the planners and engineers and extol the benefits likely to accrue from the new bridge. The crowd was said to have cheered with special gusto when the governors' wives cut ceremonial ribbons, first at the bridge's Poughkeepsie entrance and again at its Highland entrance. Ralph Modjeski, Daniel Moran, and Frederick Greene were

FIGURE 11 The Mid-Hudson Bridge at Poughkeepsie, deep into winter on the river. Photograph courtesy of the New York State Bridge Authority.

all duly recognized, and a rousing good time, complete with a marching band and a parade, appears to have been had by all. It would have been hard to tell, during the festivities, that the grim times of the Great Depression had begun and that some bad years were just ahead.[16]

Once the glow of the opening ceremonies had faded, a few signs of discontent surfaced. Construction costs had run over the original budget of $5 million, and when a local editor complained that the cost of the bridge had been "excessive," Superintendent Greene responded by attributing the problem mostly to the caisson disaster, which had required that a substantial force of state employees be kept on the payroll for an extra year or more, and by accusing the editor of political motivation.[17] The owners of the Poughkeepsie-Highland Ferry Company, which had been operating between those two locations for more than a hundred years, were justifiably apprehensive about the new bridge. It had been built virtually over their ferry slips, and within a short time two of their three boats would indeed be out of service, most of their business having vanished. About eleven years later, after drastic reductions in its rates had failed to generate enough business to justify the operation of its remaining boat, the company would shut down.[18] And it's not likely that the families of the four workers who were killed in construction accidents while working on the bridge had much cause for celebrating. Two of the men had been killed in the caisson work, and two in the superstructure work.[19]

Not much was heard from the original advocates for pedestrian access, the automobile having by 1930 pretty much taken over when it came to crossing the river, but they are unlikely to have been pleased. Although the new bridge had two 54-inch sidewalks, access to them was not easy, and satisfactory approach construction for walkers would have to wait another sixty-nine years for the state to reconfigure it.[20]

Whether or not everyone was happy about the number of vehicles crossing the bridge is not clear. It was reported that the average number of crossings during the first eight months of operation was 1,250 vehicles per day, a volume almost identical to that predicted by the Hudson Valley Bridge Association when it first advocated construction of a bridge back in 1922. On the other hand, the average daily number of vehicles traveling through the first tube of the Holland Tunnel during its first year was about 11,700.

The caisson disaster that had delayed completion took a further toll. It was the subject of a couple of lawsuits that were not finally settled for another five years. Blakeslee Rollins sued the state to recover what it said was almost $300,000 spent on righting the east caisson, claiming that the problem was the fault of the state's engineers and that the caisson, as designed, was inherently unstable. The Court of Claims didn't agree, reasoning that the design was sound and that the failure was the fault of the contractor, whose insufficiently experienced foreman (the court said) had excavated too much material from under one side of the caisson and not enough from under its other side. When the decision was appealed, the state's Appellate Court sustained the lower court's ruling.[21]

The American Bridge Company sued, too, claiming that the state had damaged it by almost $20,000 when it rejected the Bridge Company's plan to defer fabrication and shipment of its steel, forcing it to store its fabricated material off-site, handle it twice, and ultimately repaint it. One of the state's own witnesses, an old-timer who had apparently seen plenty of bridge construction, admitted in testimony that he thought American Bridge was entitled to the money, and he expressed hope that American Bridge would win the case and provide the steel on many future jobs for the state. He thought the company had performed superbly and had given the state a minimum of trouble. This time the court agreed with the claimant, awarding the full amount claimed to the Bridge Company.[22]

CHAPTER 10

The Rip Van Winkle Bridge

ONLY A FEW MONTHS AFTER THE CORNERSTONE for the Mid-Hudson Bridge had been laid at Poughkeepsie, and a full two years before the first crews actually started building anything there, plans for a bridge between the towns of Hudson and Catskill were being discussed in the New York State Legislature. It was May 1926.

A bill authorizing early studies had been introduced, and like most early efforts on such projects, it relied on as many arguments as its eager and imaginative authors could construct. It cited the unreliability of the ferry between Hudson and Athens, the periodic impact of winter ice, the convergence of multiple roads at Hudson, the high quality of the road that ran west from Catskill, the potential importance of such a bridge to truckers carrying agricultural products from areas west of the Hudson to New England, and a few other things as well. The legislature passed the bill, but Governor Al Smith vetoed it, explaining that although he thought the bridge was a good idea, he had some objections to the mechanics of this particular bill. But he added an important note to his veto, directing the superintendent of public works to go forward with whatever studies and related engineering work he could and to charge the expenses to the departmental budget. That directive would make it possible for some important engineering work to be done before all the required political and administrative issues had been resolved. None of it produced a bridge, but it did get some early thinking and planning started.[1]

There was intermittent support for the bridge over the next few years, but not until 1930 would it amount to enough to induce Representative Ellis Bentley of Wyndham, New York, to introduce a bill with some teeth in it.

By then the governor was Franklin Roosevelt, a Hudson Valley native who was known to favor almost any legislation that looked like it might serve the interests of the region. But in 1931 Roosevelt vetoed the bill, too. Like Smith, he favored the bridge but had other ideas for getting it funded. He had been impressed by the success of the Port of New York Authority, which was already building tunnels and bridges of its own, and his idea was to apply the Port Authority's model in the Hudson Valley. Toward that end, he encouraged preparation of a bill to establish the New York State Bridge Authority, which would sell bonds to raise the money needed and—like the Port Authority—would commit its bridges' revenues to amortizing and paying interest on them. Such a bill was passed and signed in 1932, and six months before Roosevelt left Albany to begin his first term as the country's thirty-second president, the New York State Bridge Authority was up and running.[2]

Considering that support for the bridge in Hudson had been limited only a few years earlier and was even less enthusiastic than what was being heard from across the river in Catskill, the votes marshaled for the 1930 legislation must have taken some of its opponents by surprise. In fact, though, success should have been predictable. The arguments that had been used to justify the original 1926 legislation had by 1930 given way to or been folded into a couple of issues that had become much more compelling as the concerns and preferences of a changing society had shifted. Now, fear of unemployment was a major concern, and travel by automobile and truck was becoming a popular preference.

As to unemployment, conditions in the Hudson Valley had by then become fairly worrisome, but they weren't yet very bad. In 1930, when unemployment had reached 8.9 percent in the country at large and averaged 6.4 percent in New York State, it was still only 5.3 percent in Greene and Columbia counties.[3] But those relatively benign conditions notwithstanding, uncertainty and fear about the economic future of the country were everywhere in 1930, with banks closing, businesses failing, and unemployment increasing, and people in the Hudson Valley were as genuinely worried as people were anywhere. Two long-established economic mainstays of the area, cement production and brickmaking, were already showing signs of weakness, and in the anxious atmosphere of the period their problems loomed large.

A big cement plant in what is now called Cementon had once been owned by the Alsen Company, a German cement manufacturer. Alsen had been identified by the War Department as an enemy-controlled business during World War I, and its plant had been closed down for a few years. It passed from one owner to another until it was ultimately acquired by the Lehigh Cement Company. By the late twenties, Lehigh hadn't yet seen enough demand to justify bringing the plant up to full capacity, and it wasn't long before two other cement companies in the area, already suffering some of the impact of the approaching depression, closed down. By the time the prospect of a bridge surfaced as a new source of local employment, the deteriorating state of affairs around Cementon was becoming a factor in the public's thinking.[4]

The picture wasn't any brighter for brick manufacturers. In the Hudson Valley they numbered 131 at the beginning of the century but had contracted to a few more than 50 by the 1920s. Annual demand had shrunk from more than a billion bricks per year in 1909 to about 200 million. By the 1930s the brick plants had begun a halting recovery, but their dominance in the huge New York City construction market was threatened by manufacturers in other parts of the country who were benefiting from newer and less expensive transportation options, from new materials, and especially from new technologies. Making matters worse, from an employment perspective, some of those competing sources were beginning to force local industry to replace outdated, labor-intensive manufacturing techniques with processes that required far fewer workers. Equipment that once produced 25,000 bricks in a day was being replaced by technology that could produce 100,000, and the size of the labor force shrank proportionately. Conditions and prospects in the brickmaking industry, which had for almost three centuries been a reliable employer in the Hudson Valley, had begun to take their toll of what remained of public optimism.[5] Under the circumstances, a new bridge that was expected to produce more than a hundred new local jobs for a couple of years had by 1930 started to look like a pretty good idea.

As to the role of automobiles and trucks as factors in the growth of support for a bridge, there's evidence that it was significant. Only twenty years earlier there had been only about 468,000 motor-driven vehicles registered in the United States, including about 10,000 trucks. By 1930, when the Bentley bill was proposed, there were about 20 million motorized vehicles, including about 3.6 million trucks. The day of the motor-driven vehicle had dawned with dazzling speed and was already profoundly changing the habits and preferences of the community, bad economic conditions notwithstanding. Earlier arguments for building a bridge had begun to look a lot more convincing, especially when they invoked the ferry's unreliability and irregularity and its vulnerability to ice, weather, and other dangers. Certainly among the new and growing community of car and truck owners, support for a bridge was by the early 1930s substantial and increasing.[6]

Public Works Superintendent Frederick Greene was a busy fellow in those days. Work on the Mid-Hudson Bridge was proceeding at breakneck speed to recover time lost to the caisson disaster, and other public works programs cluttered his desk, too. He hadn't yet done much about engineering work for a bridge at Hudson, but now the project looked likely to get going, and he had his eye out for someone to start the required design. That's when he spotted Glenn Woodruff, a senior civil engineer who had been working for Ralph Modjeski on the Mid-Hudson job in Poughkeepsie.

Woodruff, forty-two years old in 1930, had graduated from Cornell University in 1910. Until 1917 he had followed what was in those years the standard early career course that placed bright young civil engineers in the drafting rooms and on the construction sites of the country's big railroads and bridge companies. After a few years of military service during World War I, he had returned to that environment for a while and then moved into positions of increasing responsibility with such consulting firms as Robinson and Steinman, a prestigious company whose later work would include some of the country's most outstanding bridges. In 1923 he joined the eminent Modjeski firm, and in 1930 he was working as its senior engineer on the Mid-Hudson Bridge project. In that role, it was natural that he and Greene would come to know one another.[7]

By then, the toughest problems on the Mid-Hudson job had been solved, and construction of the bridge was entering its final phase. The bridge cables had been spun, and American Bridge Company was getting ready to erect the deck itself. Woodruff would later write an exceptionally clear and informative analysis of the failure of the east caisson that included an interpretation of the disaster that would be ratified by state courts of appeal.[8] With the end of the job clearly in sight, Woodruff accepted an invitation from Greene to become an independent consultant to the New York State Department of Public Works for the design of the bridge that would be called the Rip Van Winkle.

Within his first six months on the new assignment, he completed a preliminary design for what he hoped would be the actual bridge, but he was to be disappointed. His design showed a bridge that was just 240 feet less than a mile long, about two-thirds of it to be built as an elevated viaduct that would start in Hudson and carry a 30-foot-wide road and a sidewalk, first over the New York Central's tracks and then out across the marshy islands and streams that lay in the easterly part of the river. Where the viaduct reached the main shipping channel, closer to the river's west bank, Woodruff's design showed the road and sidewalk being carried across by a pair of soaring, 800-foot-long trussed steel arches, and then being continued to an intersection with the state highway at the bridge's westerly terminus in Catskill.[9]

Over a period of almost two years after the presentation of that 1930 design of Woodruff's, almost everything except his big steel arches would survive changes that naturally evolve during the preparation of working drawings. The arches, however, would disappear. Those elegant features of his design would be replaced by a through-truss cantilever system in which deep trusses supported on 125-foot-high towers would be cantilevered out into the space above the shipping channel. There, 200 feet east of the west tower and 200 feet west of the east tower, the free ends of those deep trusses would pick up the ends of a connecting truss that would close the 400-foot-gap between them.[10]

Just why the change from an arch design to a system of cantilevered trusses was made is not entirely clear, as the Bridge Authority's archive for that period is incomplete, but it seems likely that it was to save money. It's also true that being able to bridge the shipping channel with a cantilevered span that did not require falsework in the river might have played a role, but in the darkening days of the early 1930s it would have been difficult to justify the elegant arch solution when a less expensive truss option was available.

The decision to use the cantilever wasn't made easily. Only about twenty-five years had elapsed since the catastrophic 1907 failure of the cantilevered bridge that was being built across the St. Lawrence River at Quebec. That disaster killed seventy-five workers and left an indelible stain on what remained of the career of the bridge's aging, prestigious chief engineer, and it is said to have precipitated the bankruptcy of the bridge contractor as well. The Quebec failure had a chilling effect on the practice of civil engineering for a long time, influencing the thinking of bridge designers charged with choosing among options for river spans and inhibiting their creativity. Not surprisingly, the cantilever option lost a good deal of favor after Quebec.[11]

Only four other cantilevered bridges are known to have been built in North America and Europe in the dozen years after Quebec, but later, as systemic weaknesses in the relevant design and construction processes were addressed, engineers slowly began to return to the cantilever. Between 1919 and 1930 at least eleven major cantilevered bridges were built, including the George Rogers Clark Memorial Bridge on the Ohio River in Kentucky and the John P. Grace Bridge on the Cooper River in South Carolina, each with cantilevered spans of almost 1,100 feet. Closer to home for Woodruff and the others working on the Rip Van Winkle design, New York's Port Authority had by then built the Goethals Bridge and the Outerbridge Crossing, with cantilevered spans of 672 feet and 750 feet, respectively.[12] Whether Woodruff's decision to shift to the cantilever was in fact determined entirely by cost constraints, as has been thought, or whether he had by 1931 begun to feel that it was time to give the cantilever another chance is not really apparent from available records, but it is interesting to note that he would again turn to the cantilever on the assignment he took after the Rip Van Winkle.

Design and engineering were of course central to the state's effort to get a bridge underway, but they weren't the only challenges. Raising money for construction was another, and by the middle of 1932 it was time for the newly established New York State Bridge Authority to issue and sell its bridge bonds. For that purpose, it turned to the Reconstruction Finance Corporation (RFC), a federal agency that had been established by President Herbert Hoover at the beginning of his last year in office to provide financing for banks and other financial institutions. By July 1932 the RFC had authorization to extend its activities to include loans to toll bridges, and the Rip Van Winkle is said to have been the first of the bridges it considered.

Fortified with a cost estimate that had probably been prepared when Woodruff's arches were still the centerpiece of the design, the Bridge Authority was able to arrange with the RFC for the early 1933 purchase of $3 million worth of its bonds. The amount was less than the $3.4 million that had been authorized in the legislation that had established the Bridge Authority, but it was still substantially more than enough (it later turned out) to build the Rip Van Winkle. The bridge's toll revenues were pledged as collateral for the loan, and the RFC a few years later sold the bonds to the State of New York.[13]

With design work well advanced and financing assured, things had been moving ahead smoothly, but not entirely without problems. An especially nettlesome one surfaced in the process of acquiring neighboring land required for building abutments, approach roads, and the like. A property on the west bank that the Bridge Authority needed belonged to the heirs of Thomas Cole, the nineteenth-century painter who is considered to be the father of the Hudson River School of art. A dispute about how severely the Cole property would be affected by construction of the bridge was finally resolved when Department of Public Works Superintendent Frederick Greene intervened personally, arranging for design changes that obviated the taking of any of the Cole buildings (several of which would later be designated as national landmarks) and negotiating an acceptable price for what remained.[14]

What was probably the most difficult problem of all for the engineers working on the design took everyone by surprise. It was Glenn Woodruff's decision to leave the project around the middle of 1931. He had been invited by Charles Purcell, a distinguished West Coast engineer, to join his staff as "engineer of design" for the eight-mile-long, multi-span structure that would come to be called the San Francisco Bay Bridge. Ralph Modjeski was on the board that had selected Purcell as chief engineer for the big project, and there's reason to think that he had a hand in steering Purcell to Woodruff, his former employee, to head up its design. It was a spectacular opportunity for Woodruff, whose engineering assignments before the Rip Van Winkle had rarely risen to the level of responsibility that his ability and experience warranted. He is said to have been a quiet man, not inclined to make ambitious demands, and it is thought by some that his restrained behavior might have been related to a disabling speech impediment. Woodruff joined Purcell on the Bay Bridge project and went on from there to other prestigious assignments in the West, ultimately establishing his own consulting firm in 1951. An indication of the high regard in which he was held in the profession was his 1941 selection by the Federal Works Agency to be a member of the elite commission that would analyze the failure of the Tacoma Narrows Bridge. He was in good company on that commission: The other two members were Othmar Ammann, designer of the George Washington Bridge, and Theodore von Karman, the distinguished Hungarian-born Cal Tech academic whose insights into the complex behavior of fluids were the basis of his expertise in matters of wind-induced stress.[15]

By spring 1933 the Rip Van Winkle was one of the few bright spots in the darkening atmosphere of a deepening economic depression. On the first of June the Frederick Snare Corporation of New York submitted the lowest bid, in a field of twelve, to build the bridge. That winning bid, at about $2.3 million, was only slightly under the bid of Walsh Construction Company of Davenport, Iowa, and not far below the bids of the ten other fiercely competitive contractors who had been seeking the job. More important, Snare's low bid was almost $700,000 under the $3 million that the RFC had authorized. Snare was a well-established contracting firm with a good reputation, and the award was made quickly.[16]

Unlike the recently completed Mid-Hudson Bridge, where the state had awarded separate contracts for foundation work and superstructure work, the Rip Van Winkle would be built under a single contract for all the work. Snare's strong suit was foundation construction, so it subcontracted the steel superstructure to Harris Steel Company, a substantial New Jersey–based fabricator with a great deal of experience in high-rise building work but less experience in bridges. To survive in the depression-induced competition of the 1930s, Harris and similar firms were reaching beyond their traditional specialties in search of new business, and Harris had only recently begun to compete for bridge work. The price at which it contracted to furnish and erect the steel for Snare was very low, even by 1933 standards, and was probably a factor in getting Snare the general contract, but it didn't prevent Harris from performing entirely satisfactorily.

Frederick Snare had been in business since the beginning of the century, starting as Snare and Triest and becoming the Snare Corporation in about 1921. Snare himself, who was sixty years old in 1933, had during the late nineteenth century held a senior position in a Pennsylvania-based steel and bridge-building firm that was acquired by the J. P. Morgan Company when it established U.S. Steel in 1901. Snare then moved on to New York to start his own company. His partner in the new venture was Gustav Triest, a man just his age and a civil engineer active in bridge construction and married to the daughter of Charles Macdonald, the Rensselaer-educated civil engineer who had been one of the principal designers of the Poughkeepsie Railroad Bridge. Snare and Triest were able to start their new company with a solid combination of relevant skills and experience, good connections, and apparently enough capital to take on large and important work early along. Even during those beginning years, the company built all or parts of several major projects, including the big naval prison at Portsmouth, New Hampshire, piers for the Cunard Line in New York Harbor, and the piers and superstructure masonry for Gustav Lindenthal's historic Hell Gate Bridge across New York's East River.

From the very beginning, Frederick Snare had been caught up in the idea of doing big construction work in Central and South America, and early in the new century he had established a branch office in Havana, Cuba. There was a good deal of work to be done in Cuba after the Spanish American War, and he was able to secure contracts for much of it. Over the course of almost the next two decades, Snare and Triest virtually dominated the construction of extensive, badly needed harbor facilities in Cuba, and the company did more than a little such work elsewhere in Central and South America, too. By about 1921, when Frederick Snare and Gustav Triest decided to go their separate ways, the firm had become large and secure. After that, the Snare Corporation would go on to do a good deal more heavy construction on its own, including considerable major bridge work in and around New York and elsewhere. Snare himself, who had become board chairman in 1927, didn't limit his interests and energies to construction. When he died at eighty-three in Havana, he had twice been senior golf champion of the United States and had been captain of the United States Senior Golf Association team that competed regularly against a similar team in England.[17]

Snare's forces started work on the Rip Van Winkle in July 1933. Bridges are by their nature complex, vulnerable to weather, and often fraught with unanticipated problems during their construction. But in comparison to other bridges of its size, this one promised to be less difficult than most. The river's navigable channel at Catskill is adjacent to its west bank and is about 800 feet wide. It is separated from a narrow channel that flows along its east bank by a broad, uninhabited, marshy island about 1,600 feet wide. The bridge had been designed to carry a 30-foot-wide roadway and a single 54-inch sidewalk across both channels and across the island, from Hudson to Catskill.

There wasn't much in the foundation work that Snare hadn't done many times before, and by November 1933 the work had acquired some momentum. There were sixteen deep piers to be built, including the two abutments. Four of the pier excavations were to be taken down to hardpan, five to solid rock, and the other seven, which were to be supported by concrete-filled tubular steel piles driven to rock, were shallow. Things worked out about as expected, and only one of the piers proved to be especially problematic. Soundings for the most easterly pier of the cantilevered section had revealed a steeply sloping bottom, and because the engineers wanted to be able to inspect the starting surface at the very bottom of the pier hole, they required that Snare use a pneumatic caisson. Concern that it might be difficult to lower and land the caisson in a truly vertical position before starting to drive it down led to the application of a special technique. With the river bed about 20 feet below the surface of the water, Snare started work at that elevation by depositing about 10 feet of sand fill there to produce a suitably level starting surface for landing the caisson. Then a braced, sheet-pile cofferdam with a footprint larger than the caisson's was driven down into and through the newly placed fill and dewatered. Once the caisson's cutting edge had been lowered into position within the cofferdam, successive lifts of concrete were placed above it while its interior was dredged, causing the whole structure to be forced down by its increasing weight deep into the river bottom. Once it had reached hardpan, 90 feet below the water's surface, the engineers were able to satisfy themselves that they had secured adequate support and that no further concrete was needed. The caisson's working chamber and its dredging wells were pumped out and filled with concrete to produce the desired massive pier, ready to support a masonry pedestal and a steel tower for the new bridge.[18]

Mixing, distributing, and placing concrete in all those deep pier holes was probably a bit easier than the difficult and often hazardous work of digging, dewatering, and otherwise preparing the holes, but not by much. The piers were stretched out along a line that followed the centerline of the mile-long bridge, a few of them in shallow water and the rest in marshy terrain that denied access to all but the most tenacious equipment. Snare mixed most of the concrete on barges out in the river, transferring it by derrick to a big shore-based hopper. From there, crews distributed it in bottom-dump buckets carried out to the pier sites on a temporary narrow-gauge railroad system that Snare had built along the boggy surface of the island.[19]

Winter on the Hudson came early in 1933 and was one of the coldest on record, shortening Snare's first working season and slowing progress a little, but by May 1934 many of the piers and some of the masonry pedestals were in place, and Harris Steel had arrived at the site and started to erect its first steel towers. Some of the bridge's 12,000 tons of steel were already reaching the site on a temporary narrow-gauge railroad spur, and some would soon arrive on barges via Catskill Creek. Snare's temporary railroad was probably used for distributing some of the viaduct steel, but the massive truss sections that would form the main cantilevers would be placed on their own barges and lifted into position by traveler cranes operating from the growing structure above.

By the end of November eleven of the spans were in place, and a month later all hands turned to the formidable task of finishing up the big truss work above the shipping channel. The 1934 winter was proving to be almost as bad as the previous one, but about sixty ironworkers would spend much of a cold January almost 200 feet above the river, fighting a bitter north wind for control of the last big sections of steel. By the eighteenth of the month they had coaxed the main truss into place, completing a connection between the ends of the east and west cantilevers and making the Rip Van Winkle whole. By the end of January 1935 everything had been bolted up and riveted, and the topping-out flag had been raised.[20]

Plans were by then in the works for a July opening, but there was still plenty of work to do. The Dutch Colonial–style administration building that would become the first permanent headquarters of the Bridge Authority, at the Greene County end of the bridge, was well along, but there was still other work to be done if everything was going to be up and running by summer. In addition, a mile of concrete pavement needed to be put in place on the bridge deck itself, and all the steel had to be given a final field

FIGURE 12 The Rip Van Winkle Bridge, between Catskill and Hudson. Photograph courtesy of the New York State Bridge Authority.

coat of paint. Dozens of other, smaller jobs remained to be done, including the installation of the bridge lighting system and even some landscaping work at the approaches.

But of course—when compared with what had already been accomplished—those were the least difficult parts of the job, and by the first of July 1935 the bridge was finished, its fresh coat of silver paint gleaming in the splendid summer weather. The gloom of the Depression was hard to find when the bridge was opened with joyous ceremony the following day, complete with flags, bands, floats, and a full complement of nattily dressed dignitaries. Governor Herbert Lehman dedicated the new bridge with a brief and informal speech that commended everyone who had a part in building it, and he predicted good things for the towns it connected. A motorcade of fifty cars demonstrated the bridge's reliability by driving across it, and an estimated seven thousand people cheered the cutting of ribbons by Colonel Greene's wife at both bridge entrances.[21]

A year later there might have been some disappointment that traffic over the Rip Van Winkle during its first twelve months of operation amounted to only about 200,000 vehicles, a smaller volume than anticipated and one that didn't augur well for the bridge's ability to pay for its maintenance and to meet its bond obligations. But there was no real need to worry. The legislation that had established the New York State Bridge Authority had intentionally avoided limiting its authority to the Rip Van Winkle alone. In fact, it had always been Roosevelt's intention that the Bridge Authority, like the Port of New York Authority, would build or absorb other bridges in the region and that the revenues of all such bridges would be pooled, as a way of spreading the risks and buttressing the financial strength of weaker performers. Before he left the governor's office in 1932, FDR had seen to it that ownership and responsibility for operating the prospering Mid-Hudson Bridge at Poughkeepsie, which carried about 400,000 vehicles in its first year, had been transferred to the Bridge Authority.[22]

CHAPTER 11

The Tappan Zee Bridge

By most accounts, conditions in America during the late 1920s were about as good as they'd ever been, and it was hard for most people to accept the idea that things might not always be that way. World War I had ended about a decade earlier, new businesses were starting almost everywhere, and decent jobs were abundant. The automobile was beginning to provide families a mobility that promised to broaden their housing and employment options and enrich their lives. Along the lower Hudson, the Bear Mountain Bridge was up and running, the Holland Tunnel had just been completed, the George Washington Bridge was under construction, and planning had been started for building a second tunnel between New York and New Jersey.

The proprietor of the ferry that had been operating between Tarrytown and Nyack for almost a hundred years is unlikely to have shared in all that optimistic thinking. Positioned about eighteen miles north of the George Washington Bridge and about as many miles south of the Bear Mountain Bridge, he hadn't yet lost a significant amount of ferry traffic to either of those new crossings. But he had recently been hearing and reading about a growing movement to build still another bridge, and this one was going to cross the river between Rockland County, on the west bank of the Hudson, and Westchester County, on the east bank. Such a bridge would be right on his doorstep, and it could reasonably have been a source of concern to him. As things would later play out, it should have been.

In the 1920s support for such a bridge was limited. Rockland County was still largely rural and sparsely settled, and although many of its agricultural products were being trucked to New York, the George Washington

Bridge provided all the access it needed. Westchester was growing, but its interests lay mainly south, in New York, and there wasn't yet much justification or enthusiasm for a bridge to Rockland. But by the early 1930s, despite the depressed state of the economy, voices favoring such a bridge were becoming louder. A Rockland County legislator named Ferdinand Horn rallied some support for a bridge between Piermont and Irvington, across a reach of the river only a short distance south of the Nyack-Tarrytown line, and in 1932 he introduced (but failed to get ratification for) a bill to authorize its construction. A few years of debate followed, and by 1935 the legislature had become convinced of the idea's merit and established the Rockland-Westchester Bridge Authority. For the first time, a bridge between the counties of Rockland and Westchester began to look like a real possibility.[1]

Such modestly increasing support, combined with a measure of legislative approval, brought out the opposition in force for the first time. But before its leaders were able to organize much resistance, the bridge had become a casualty without their help. It was revealed that the bistate agreement that back in 1921 had established the Port of New York Authority (later called the Port Authority of New York and New Jersey) had conferred on that agency exclusive responsibility for building and operating all Hudson River bridges and tunnels south of a line drawn between Nyack and Tarrytown. The simple truth was that no one but the Port Authority could build a bridge between Piermont and Irvington. That was bad news for the bridge supporters, who didn't have anything against the Port Authority but were loath to lose control of the revenues they expected from their bridge. The rule was in fact entirely reasonable. It had been designed to prevent private-sector speculators and others from building bridges or tunnels that would compete for revenues with (and undermine the economic stability of) what the Port Authority expected to be building.

That bad news about crossing at Piermont didn't stop the bridge boosters. They regrouped and renewed their campaign with a proposal to build north of the restricted zone instead, and late in 1935 they replaced the first bridge authority with the Tarrytown-Nyack Bridge Authority and engaged engineers to do preliminary studies and take borings along a route that would directly connect those two towns. The opposition didn't quit, either, and it was able to maintain a high profile in the press and elsewhere with continuing anti-bridge pronouncements from local historical societies, conservation groups, and prominent citizens, including one from the head of the mathematics department at Stevens Institute of Technology.

The issue attracted plenty of attention, but between the burdens of limited funds and economically depressed conditions, a bridge between Tarrytown and Nyack never really had much of a chance in the 1930s. The second bridge campaign ended in November 1936, when the engineers reported that the borings had failed to find adequate bearing at depths less than almost 200 feet below the water's surface, and the Bridge Authority reasoned that such a condition would make the cost of a bridge higher than anything that could be justified.[2]

Occasional efforts to renew the bridge campaign continued to surface for a few years, waning as the impact of the Depression deepened and disappearing entirely from public discourse when World War II started. But the idea of a bridge at Tarrytown hadn't entirely disappeared from the public's thinking, and other events that couldn't have been seen as having any connection with a bridge were already unfolding in unlikely places.

The star of Robert Moses had been rising in New York ever since Governor Al Smith appointed that often controversial visionary to the presidency of the Long Island Park Commission in 1924, a gesture that significantly enhanced a career trajectory that would eventually make Moses one of the most powerful men in New York. He was admired by some for his brilliance as a master planner and hated by others for his arrogance and for what was seen as his indifference toward the people who were most directly affected by what he built. Under his leadership, the Long Island commission and many other agencies became the builders of roads, bridges, and parks of such vastness that they would change the region forever.

In 1927, when his still-young Long Island commission was building a kind of American Riviera that would become the central feature of Jones Beach State Park, the contractor on one of its larger projects went broke. In the negotiations to ensure that the work would continue without prolonged interruption, Moses found himself working closely with the failed contractor's superintendent, a thirty-three-year-old, rough-edged Irishman named Michael John Madigan, usually called "Jack." According to Robert Caro's masterly biography of Moses, Moses was quickly impressed by young Madigan's practical knowledge of the project, by the clarity of his ideas about what should be done, and especially by his restrained poise in presenting and defending his ideas. The perceptive Moses developed a friendship with

Madigan that would last for a long time and would become a powerful force in the young superintendent's career.

Jack Madigan was born in Danbury, Connecticut, in 1894 and had abandoned home and school by the time he was thirteen to begin working as a waterboy on construction sites. It took fewer than about ten years for him to learn enough about construction to get himself work as a labor foreman on small jobs, and within only a few more years he was running big jobs. Smart and ambitious, he became a construction superintendent in his early thirties and began to acquire some of the commercial and social skills that would not many years later make him a rich and influential figure in New York.

About a year after the start of his friendship with Robert Moses, Madigan left construction to join forces with Richard V. Hyland, a Notre Dame–educated civil engineer, to establish Madigan Hyland, Consulting Engineers, with a base in Long Island City. Hyland brought the engineering credentials, and Madigan provided everything else. It was a relatively low-risk venture, probably underfunded, and, at least for a while, understaffed, but with Robert Moses as its patron, those would be modest and temporary burdens. Caro writes that Moses "gave them Long Island Park Commission contracts . . . that made Madigan a millionaire." The young company went on to develop a staff of well-educated engineers. When asked about his own professional credentials, Madigan responded (according to Caro) that he regarded his staff as his "Phi Beta Kappa keys." Caro's descriptions and references to the relationship between Madigan and Moses, as it played out over the years, suggest that Moses attached great value to the wisdom and practical savvy that Madigan brought to the friendship, his rough edges notwithstanding.[3]

What Madigan might have lacked in engineering skills he made up for in his ability to pick good engineers, and within a few years of establishing the Madigan Hyland firm, he had hired the innovative and often brilliant Emil Praeger as his chief engineer. Praeger had graduated from Rensselaer (where he had been a varsity football player) in 1915, and after entry-level employment in civil engineering work and then service in the navy during World War I, he had returned to New York, where he turned his attention to the structural design of buildings. For most of the 1920s, he worked as principal engineer in the distinguished New York architectural firm of Bertram Goodhue, doing engineering work for monumental (mostly reinforced concrete) institutional buildings the firm was designing in such widely separated locations as California, Hawaii, Illinois, Indiana, Nebraska, and New York. In California alone, during Praeger's years with the firm, his work included structural engineering for eight or nine buildings at the California Institute of Technology and for the historic Central Branch Building of the Los Angeles Public Library.

When Goodhue died unexpectedly at the age of fifty-five, Praeger elected to remain with his successor firm for a few years, but in 1929 he made a radical change, signing on as assistant chief engineer of the Curtiss-Wright Corporation, the country's largest manufacturer of airplanes. By that time he had developed a solid reputation as a capable engineer whose talents and skills were centered mainly in large structures, and he probably found the new job as assistant to the exceptionally dynamic chief engineer of a large aircraft manufacturing company more problematic than he had expected. Eventually, Praeger himself was made chief engineer, but that role apparently didn't entirely please him. When Jack Madigan spotted him and offered him the job of chief engineer for the Madigan Hyland firm, with responsibility for designing roads, bridges, and the like, Praeger accepted.[4]

It was then about 1934, around the time that Robert Moses, who had by then added the powerful New York City Department of Parks to his abundant responsibilities, was getting ready to authorize the start of design work for the arch bridge that would carry the Henry Hudson Parkway across the Harlem River. That historic bridge would mark the beginning of a welcome new direction for Praeger, who would have a senior role (along with others) in its design and would go on from there to become a dominant figure in Madigan Hyland.

He plunged into his work, relishing the immense variety of engineering projects generated by the Moses agenda, which were supplemented by a healthy range of other jobs as well. But it would all be cut short by World War II. Praeger was commissioned an officer in the U.S. Navy's Civil Engineer Corps (his second hitch), with responsibility for managing the design and construction of some of the service's civilian and military facilities.

Not long after Praeger entered the navy, Allied commanders in England began to plan for the invasion of Europe, and early in 1943 they had begun to search for a harbor on the coast of France that could accommodate the vast armada they planned to launch. What they were looking for was a place capable of handling more than 10,000 tons of material and equipment per day, starting a few days after the first invading troops had been put ashore. They figured it would have to be about the size of the British harbor at

Dover, which had (in peacetime) taken seven years to build, and the pickings on the French coast looked slim. Cherbourg and Le Havre had ideal harbors, but the Germans had such a tight grip on both that the Allied commanders rejected them. Instead, they made a controversial decision to fabricate their own harbor secretly in England and to tow its components across the Channel to the coast of France as soon as the invasion got underway. The task was seen by many (including some American commanders) as impossible, but it was favored by some of the top brass in the combined forces of the United States and England and by Winston Churchill personally, so the plan had what it took to become authorized.

In the approved plan there would be two harbors, one for the Americans at Omaha Beach and the other for the British at Gold Beach. Each harbor would comprise approximately the same elements and would be surrounded by a system of massive concrete breakwaters designed to protect areas where vehicles, equipment, and supplies would be unloaded and transferred to the beaches. The breakwaters—huge, buoyant concrete caissons called Phoenixes—would be anchored to the seabed. Gaps between and around them would be filled with the sunken hulls of vessels that had been brought across the Channel under their own steam and then scuttled. The docks within the protected areas would comprise additional anchored caissons that would rise and fall with the tide and be connected to the French beaches by pontoon-supported, flexible steel roadways. The components of such a manmade harbor would have to be designed and fabricated over the course of a little over a year by thousands of engineers and workers in dozens of secret locations that were (or could be) connected to the sea by water. When built, these gigantic concrete boxes would be towed with all possible stealth across the water to the south coast of England, where they would be sunk and later refloated at the time of the invasion.

At the time that work on the harbor project was getting underway, Praeger was a three-stripe navy commander, and he was given one of the senior roles in the group that was designing and building the reinforced concrete Phoenixes. A total of 147 would be needed, most of them about 250 feet long by as much as 60 feet wide and more than 35 feet high, partitioned off by concrete walls to allow control of their buoyancy. They would weigh between about 2,000 and 6,000 tons apiece. The work of designing and fabricating all the required components moved steadily forward, and the finished products were unobtrusively towed to the south coast of England and hidden from view, as planned. A few days after the invasion was launched on 6 June 1944, a fleet of British and American tugs began towing the floating components of the two harbors across the English Channel at an average speed of about four knots, delivering them to the American Seabees and to their counterparts in the Royal Navy to assemble and put them in place. About a week later, the first vessels began arriving from England and unloading supplies and equipment at the two new harbors for shipment to troops that had moved on to interior positions in France. The plan had worked.[5]

D-Day didn't end the war, of course, but within another year it would be all over in Europe. Praeger would return to New York an even more experienced and capable engineer than he had left it, and not a day too soon for Jack Madigan, who had plenty of work waiting for him and the prospect of more to come. For starters, Madigan had a job that would seem a bit like an extension of what Praeger had been doing in Europe. A devastating fire had destroyed Pier 57, a huge structure that had reached out into the Hudson River at Fifteenth Street in New York and had been used for years by the Grace Line for mooring its vessels and holding its passengers and cargoes. The city, which controlled the property and leased it to tenants, wanted the pier replaced as quickly as possible, but its engineers made one thing clear early along: This time they didn't want timber piles. The 4,400 timber piles that had supported the original structure had long since been weakened by marine borers and were thought to have provided the principal fuel for the fire that brought down the pier.

When Praeger learned that wood piles were unacceptable, he did some calculations for concrete or steel piles, and when he found that they would be too expensive, he began to think about buoyant concrete boxes like the ones that had been built for the Normandy invasion. Using such boxes to carry about 90 percent of the load and then driving steel and concrete piles through them and down to firm bearing to pick up the other 10 percent looked like a workable and economical solution. He developed such a design, the city's engineers accepted it, and Praeger was back in the buoyant concrete box business. It was Normandy all over again: Three gigantic, buoyant boxes would be built at a location far up the Hudson, where there was enough room to do it, then towed down the river to Fifteenth Street, and finally sunk there. It wouldn't be quite that simple, of course, but that was the general outline of the approach, and it would all work well.[6]

It would be a few years before the Pier 57 job moved ahead, and when it did, there were at least a couple of notable differences between what Praeger had experienced in Europe and what he was doing in New York. One difference was the welcome absence of secrecy. In fact, the whole process became something of a public spectacle, with engineering students visiting the casting site and locals along the Hudson lining the riverbanks to see the giant boxes being towed downstream. The other difference was less benign. There had been no cost constraints in England, where the survival of the British Empire was at stake, but in New York there was a budget that was inviolable. When the bids came in, they were much higher than Praeger's estimates, and the city had to scurry furiously to find approximately $2 million more to get the work done.

While all this was going on, Governor Thomas E. Dewey had been talking with Jack Madigan about matters that would be central to life at Madigan Hyland for more than a few of the years ahead. Dewey had for some time been nurturing a plan to build a superhighway across the entire state of New York, from Buffalo to New York City. He saw it as a latter-day Erie Canal, except that the canal had terminated on the west bank of the Hudson and Dewey's highway would have to cross the river on a grand bridge. After many years, a bridge between Rockland County and Westchester County might have its first serious constituency, starting with the governor of the state.

With the size and mobility of the postwar population growing rapidly, pressure to get on with planning and building Dewey's superhighway increased quickly. He had already made a few halting efforts to get it started by using state funds, but he had become convinced that what he wanted would never be built unless it could be made self-supporting. In 1949 he convened a high-powered committee to explore such an approach, and its members agreed with and supported the idea that whatever was built should be paid for by the people who used it. They laid out guidelines for the necessary legislation, and by March 1950 Dewey had signed the New York State Thruway Act into law. It established the New York State Thruway Authority and empowered it to borrow the necessary money to design and build the roads, bridges, and other infrastructure and to collect tolls sufficient to make the whole enterprise self-supporting. Later, the legislature would follow the leads of several other states by proposing the use of state credit to back the thruway's bonds, and once the electorate ratified such a proposal by referendum, the authority's borrowing costs would become extremely low. Dewey put the state's superintendent of public works, Bertram Tallamy, in charge of the new authority, and by September it had sold its first $10 million worth of bonds and was starting to enter into contracts with engineers all over the state to design the system.

There would be plenty of work for Madigan Hyland in this grand enterprise, and in fact the company was already well along on some of it: a survey of existing traffic along the proposed route and estimates of the volume of future traffic. That study was of course the critical first step in the design of what would become an approximately 550-mile highway, but it was only a modest prelude to what would follow for Madigan Hyland: a commission to design the approximately 3-mile-long bridge that would carry the new highway across the Hudson between Nyack and Tarrytown.[7]

For Praeger, who was fifty-six years old in 1950, the Hudson River span would be the first major bridge on which he would have ultimate design responsibility. There had been bridges in his work before, during the Moses years, but none on the scale of what would come to be called the Tappan Zee. He plunged into preparing a preliminary design while Jack Madigan and the governor were still talking. If the former captain felt a twinge of nostalgia for the secrecy that had characterized the Normandy work, it wasn't for long. Governor Dewey, aware that there were still passionate advocates for different locations for the bridge and even a few lingering opponents who favored abandoning the bridge plan entirely, imposed a pattern of secrecy that would last almost a year. He wanted to give Praeger and his staff a chance to put together a design without having to deal with the dissenters.

As it turned out, not much more would be heard from the groups that opposed any bridge at all, but there was plenty of vigorous protest from advocates of alternative locations. When word leaked out that Praeger's scheme would include an approach section through the prosperous hamlet of Grand View, just south of Nyack, displacing an established and voluble population of artists and writers, the press gave the protest a high profile. Then the Port Authority came along with a plan to build its own bridge at Dobbs Ferry, just a few miles south of Tarrytown and well within the Port Authority's assigned territory. That plan was seen as a real threat to Dewey's bridge, and it showed signs of having originated with or at least having received enthusiastic support from Governor Alfred Driscoll of New Jersey, who favored any location that would allow easier connection to highways that New Jersey was building or planning. But Dewey, a tough-minded former New York City district attorney, bridled at the Port Authority's idea and worked out a solution with Driscoll that made short work of the Dobbs Ferry option.[8]

The opposition retreated, but it would be a while before it surrendered. One argument that gained enough support to slow things down a bit held that the bridge's location at what was perceived as the river's widest place was an expensive error that should be corrected. In fact, the location was only the second-widest, but that was a niggling fact of little real consequence. The objection had enough substance to gain some important support, but the Thruway Authority stood its ground. Chairman Tallamy explained that the north and south limits of the route had been effectively determined by the already carefully established location of the thruway itself, and that although some flexibility remained, it wasn't much. North of the selected route, he said, the topography of the west side of the river was problematic and would require some very expensive grading and construction. In addition, he argued, a more northerly crossing would require the acquisition of a considerable amount of extraordinarily expensive real estate on the river's east bank, in the densely populated, upscale county of Westchester. Starting about 800 feet south of the selected route, the territorial jurisdiction of the Port Authority denied the Thruway Authority the right to cross.

While all this wrangling and debating continued and eventually receded, Praeger and his staff at Madigan Hyland were busy designing the bridge. Extensive new borings confirmed earlier probes that showed subsurface conditions varying widely along the proposed route. Hard rock showed up near the surface of the ground at the eastern shore, but that was about the extent of the good news. Not far west of the eastern edge of the river the rock elevation fell off sharply, and at mid-span it was down 300 feet. By the time the boring crews got to the west shore, they had to drive their probes more than 1,400 feet to find rock.

What emerged late in 1950 was a scheme showing a bridge almost 16,000 feet long with seven lanes, including a median strip with an open grating in the center. Essentially, the proposed bridge consisted of two approach sections, one from the east shore and one from the west shore, connected by a section that would leap across the approximately 1,200-foot-wide shipping channel. The eastern approach section, which would start about a mile south of the village of Tarrytown, was about 3,000 feet long and was to be carried by twelve pairs of "deck girders," deep steel trusses about 250 feet long, which would support the roadway from below. At the bridge's western end, in South Nyack, the first approach section of the bridge was at that time shown as an earth-and-stone-filled causeway that would extend about a mile and a half into the river before beginning its rise as an elevated deck truss structure like the one proposed for the eastern approach. Those two approach systems, in Praeger's original scheme, were to be connected to one another by a pair of soaring tied arches that would carry the thruway across the channel at an elevation that would leave about 135 feet of clearance above the water.

The two bridge approaches would be carried on pile-supported foundations. Massive pile caps supported by end-bearing steel piles were designed for the eastern approach, where suitable supporting strata could be found at acceptable depths, but pile caps supported by large clusters of very long, friction-dependent timber piles would be needed for the entire western approach, where suitable end bearing was simply out of reach. For the center span, where the loads imposed by the long span would be tremendously greater, Praeger returned for the second time to his Normandy experience. Once again, he would rely on gigantic buoyant concrete boxes that would be cast in a remote location, towed down the river, and then positioned along the central section of the bridge route to carry about half the load of the main span, leaving it to a relatively small number of very large concrete-filled pipe piles to pick up the other half of the load.

When Praeger revealed his design, he acknowledged that it included some techniques that hadn't been used in the construction of most earlier bridges, but he was at pains to emphasize that these represented nothing more than well-accepted engineering principles that were being applied in sometimes new ways. The commissioners of the Thruway Authority fully supported him and said nothing to undermine his defense of his design, but they—not unlike most public officials with such major responsibilities—sought a complete, independent review of the design from an eminent board of engineers that included, among others, David Steinman and Carlton Proctor. Years earlier, Steinman had been an important figure (along with Othmar Ammann) in the design of Gustav Lindenthal's Hell Gate Bridge, and he had since designed some of the world's longest and most spectacular bridges. Proctor, a distinguished engineer who specialized in soils and related subsurface matters, was Daniel Moran's partner by then and would in 1952 become president of the American Society of Civil Engineers. In a unanimous endorsement of Praeger's design, the board said that it saw nothing that would prevent implementation at reasonable cost.[9]

With preliminary drawings well advanced and public opinion beginning (grudgingly) to accept the scheme, there was by the end of 1950 a rising level of optimism about getting underway with the preparation of construction documents and, ultimately, with construction of the bridge itself. But there were still a few issues that had to be resolved.

One of those issues had been created by a restrictive provision in the legislation that had in 1932 established the New York State Bridge Authority. It was something the planners hadn't counted on, and it had nothing to do with the 1921 Port Authority compact, which they had already satisfactorily addressed. The new problem had been created by a provision in the Bridge Authority's enabling legislation that was intended to protect the bondholders of Hudson River bridges that existed in 1932. It promised not to allow any other bridges to be built south of an imaginary line drawn about fifteen miles north of the Rip Van Winkle Bridge, which connected Catskill with Hudson across a distant, upstate section of the river. Enforcement of that restriction would have meant the end of the Tarrytown-Nyack bridge plan, but some skillful negotiation between the state administration at Albany and representatives of the bondholders produced an agreement that cleared the way for a bridge between Nyack and Tarrytown. In March 1951, the legislation having been changed, the state legislature gave its approval for the new bridge.[10]

A second potential impasse was more complicated and required a fairly radical change in the bridge's design. The Army Corps of Engineers must sign off on any bridge across a navigable waterway before construction can be started, and the corps didn't like that long causeway between the western shoreline and the beginning of the elevated section of the approach to the bridge. The principal concern was that ice floes blocked by the solid causeway might accumulate and become a threat to navigation. There's some evidence that Praeger might have anticipated such an objection, because there's no indication that he protested it, and he was able to produce an acceptable alternative design very quickly, substituting an elevated, essentially level (but more expensive) trestle for the earth-supported causeway. By May 1951 the corps had followed the state legislature in approving construction of the new bridge.[11]

The revised approach for the elevated trestle added about 12,000 tons of steel to the 48,000 tons previously estimated. Even after allowing for cost savings generated by eliminating the earth-and-stone causeway, the estimated total cost of the bridge was now getting close to $60 million. But by the spring months of 1951 there was such widespread enthusiasm about moving ahead with the bridge and completing it by 1954, to coordinate with the thruway's construction schedule, that within a few months the project was moving ahead at full speed.

The first order of business was to convert the preliminary foundation design to a set of construction documents suitable for bidding. It would be misleading to characterize any part of the bridge's foundation work as simple or easy, but it's probably fair to say that building the foundations for the approximately 1.5-mile-long trestle was less difficult than building the foundations for the rest of the bridge. Praeger started his construction documents with those trestle foundations, and by the beginning of 1952 the drawings were complete and ready for bidders. Construction Aggregates Inc., a New York–based heavy construction contractor, submitted the winning bid in February, and by June it had begun to convert what had once been a quiet boatyard on the Nyack waterfront to a vast and busy storage and equipment yard. The approximately 20,000 timber piles that would be needed, too long (at about 90 feet) for single railroad cars, were brought from the West Coast on tandem cars and soon began to show up at a nearby rail depot. At about the same time, the first loads of cement, sand, and stone for the 154 pile caps and concrete bents began arriving by barge at the Nyack dock. The yard itself was soon alive with cranes, mixers, and other equipment, and by September almost three hundred men had been recruited and were preparing to put in place as much work as possible before winter.[12]

All that preparation was only for the smallest, least difficult part of the bridge's foundations. The more difficult work, the foundations and concrete structures that would be built between the east end of the trestle and the east bank of the river, would be the subject of a separate, much larger contract that had not yet been awarded.

To carry the nineteen approximately 250-foot-long deck truss sections that flanked the main channel section of the bridge, Praeger's design required well-proven but extremely demanding construction: an array of high, two-leg concrete bents on massive cylindrical concrete piers supported by long steel piles driven down to rock. The arduous and often treacherous work of excavating and dewatering the 60-foot-deep holes required for the piers, then driving piles that would be started at the bottoms of the pier holes, and then building the big piers themselves and the high concrete bents that would tower above them would have been major challenges on any bridge. On the Nyack-Tarrytown job, even those tasks paled a bit when compared with what Praeger had planned for carrying the big loads imposed at and immediately adjacent to the main span itself.

There, he would once again invoke ideas that he and others had first applied in the harbor at Normandy and that he had revealed in his preliminary drawings. Eight giant floating concrete boxes would carry about half of the big load of the main span, two of them about 190 feet long and all of them at least 100 feet wide by about 40 feet high, a little smaller than some of the earlier boxes but still big enough to challenge even the most skilled contractor. Concrete-filled steel piles, some with steel columns embedded in them, would be driven through sleeves in the boxes down to firm bearing almost 300 feet below. Unlike the boxes in the Normandy system, these would have to be designed to survive for many years, with more refined detailing, better concrete, and more effective watertightness.[13]

One advantage Praeger had in 1952, when the New York State Department of Public Works invited bids for the big foundation contract, was the unique prior experience of some of the bidders. Merritt-Chapman & Scott, a big New York–based firm that had started in the marine salvage business in the middle of the nineteenth century and had expanded into the marine construction business, had only two years earlier taken the contract for the big boxes that Praeger had designed for Pier 57.[14] Corbetta Construction Company, the preeminent concrete contractor in the New York area, had been part of a joint venture that had come in second on that bidding. After the Pier 57 contract was awarded, Merritt-Chapman had subcontracted the fabrication of the floating concrete boxes to Corbetta, and the two firms had gone on to work together, accumulating some valuable expertise in the process.

As their work site for the Pier 57 job, Corbetta and Merritt-Chapman & Scott had taken over an abandoned claypit the size of six football fields along the western shore of the Hudson, in Haverstraw, and after converting it to a drydock, they had fabricated the boxes inside it. By September 1952, when invitations were issued for bids on the big foundation contract for the new bridge at Tarrytown, Merritt-Chapman and Corbetta had finished building their Pier 57 boxes, had towed them down the Hudson to Fifteenth Street in New York, and were in the process of sinking them. Their intention had been to shut down the Haverstraw site as soon as they finished up at Pier 57, but the new job, only twelve miles south of the drydock, changed everything. Once they learned the details of the bridge's buoyant foundation design, the two contractors recognized one another's unique qualifications for the job and decided to join forces again. This time they would bid as a single venture, with Merritt-Chapman & Scott as the prime contractor and with Corbetta as its subcontractor for the concrete work. There probably weren't two companies in the world better positioned to do what it would take to implement Praeger's design for the new bridge.

Not surprisingly, Merritt-Chapman's bid of just under $12 million for the foundations was the lowest received, and it was quickly accepted. By the end of 1952 a contract was in place, and the 1952–1953 winter provided some needed time to organize for the difficult first task of reopening and restoring the recently abandoned casting yard at Haverstraw. For the Pier 57 job, the companies had built a long, broad seawall that looked and functioned like a giant levee along the river's edge, then dewatered the big claypit on the land side of the wall and covered its floor with a smooth concrete slab to provide a base for casting the Pier 57 boxes. When that casting work ended early in 1952, they had removed enough of the seawall to allow the basin to refill with water so that the big boxes could float out into the river and be towed down to New York. Now they would have to start all over, and it would take Merritt-Chapman about a month to restore the big seawall, dewater the basin again, and scrape out all the mud and slime that had accumulated on the surface of the slab. By spring the site was ready for Corbetta to begin building boxes again.

It was like building eight tremendous ships. The largest of these boxes would weigh about 23,000 tons, as much as six navy destroyers, and everything else about the job was huge, too, from its two- and three-foot-thick walls and slabs to the 4,000 tons of heavy bars that would be needed to reinforce them. Batch plants were set up on-site to produce the required 32,000 cubic yards of concrete. If the schedule of the bridge construction was to be met, Corbetta figured it had about six months to build the boxes and tow them downriver in time for Merritt-Chapman to put them in place.[15] Down at the bridge site, where Merritt-Chapman was already working on the foundations for the approach structures, there would still be plenty of work to do when the boxes arrived, in order to make them ready to receive the steel framing they were designed to support.

In fact, though, something critical had happened while all that preparatory work was going on at Haverstraw. Late in 1952 the state had invited bids to furnish and erect the structural steel for the bridge's superstructure, including the tied-arch section that had become the centerpiece of its distinctive profile, and it had announced the engineer's estimate of about $36 million for the job. Only American Bridge Company (a division of U.S. Steel) and Bethlehem Steel Company were big enough to do the job, and both acquired bidding documents. Within a few weeks, both companies advised the state that, in their opinions, the job could not be done for the estimated amount, and because state law would not permit an award to be made for anything exceeding the engineer's estimate, they felt they would be wasting their time if they continued to prepare bids. There's evidence that Praeger began work on an alternate design as soon as he got that bad news but that he elected to allow the invitation to stand anyway, hoping for a change of heart by at least one of the bidders. But there would be no such good luck. At the designated time for opening the bids, on December 15, 1952, there were no bids to open.[16]

That was the end of the grand tied-arch design, and it was the end of the schedule that was controlling the work, too. Praeger produced an alternate design for the cantilevered structure that would ultimately be built. In the new design, the roadway would be carried across the main span by deep trusses that would be anchored in flanking spans and cantilevered out over the shipping channel from giant steel towers on each side of the gap. It would be a less expensive, less popular, and certainly less elegant bridge, but it was one that could be built within the budget. Four months later, in April 1953, American Bridge Company would respond to the revised drawings with the winning bid to fabricate and erect the cantilever design. Its price was about $7 million lower than the state's estimate for the tied-arch scheme.[17]

By the time a contract was signed with American Bridge, Merritt-Chapman and Corbetta had finished building their boxes and had towed them to the bridge site, where workmen were well along in the process of securing them in place. Little of what the contractors had been doing had been easy. All eight boxes had been towed downstream before their top slabs had been placed and before their sidewalls had been brought up to full height, because the Hudson's shipping channel wasn't deep enough to accommodate even their unladen draft. Leaving off the tops and some of the walls saved enough weight to do the trick, and the boxes managed to clear the river bottom, but substantial work was left to be done at the bridge site. Once the boxes arrived there and had been maneuvered into position, Merritt-Chapman began the job of completing their walls and slabs. After some work had been done to prepare the river bottom below, ballast was added to sink the boxes deep enough so that only their tops showed at the water's surface. Sleeves had been left in their slabs to allow the driving of long spliced lengths of 30-inch-diameter pipe piles through them and down to deeply buried rock.

While all this was going on, American Bridge Company was busy fabricating (in five of its plants) the approximately 60,000 tons of steel that would be required for the superstructure of the bridge, and about three months into 1954 the first big fabricated sections of what it had produced were being loaded onto railroad cars destined for New York State. By then, its field managers had leased an abandoned brickyard along the river at Haverstraw, not far from Corbetta's casting site. There crews would assemble the big trusses and integrate them into even larger sections of bridge structure before shipping it all downriver to the construction site.

In June, American Bridge's Haverstraw crew built two very large steel scaffolds, each about 55 feet high, and placed them on two big barges moored at the company's dock. The scaffolds, more precisely called "falsework," would support the first of nineteen 900-ton sections of bridge structure that would be assembled at Haverstraw. Using long boom cranes and 500-ton jacks, the crew raised the first such section of bridge structure and laid it across the tops of the two falsework structures, allowing it to span between them and to overhang them by a few feet. This precipitously balanced assembly of steel, its bottom perched about 55 feet above the water, was about 250 feet long, 93 feet wide, and 28 feet high. When it had all been made secure, three tugboats towed the pair of barges with their delicately balanced cargo down to Nyack on a mercifully calm river behind a Coast Guard escort, and positioned them directly in the center of the space between two of the bridge's recently completed concrete towers, which were spaced about 250 feet apart. By timing their arrival to coincide with high tide, the crews were able to keep the bottom of the big assembly a few feet above the tops of the two towers that were waiting to support them, and as soon as the tide began to turn, the falling level of the water gently lowered the barges and their cargo. Within a few hours, the bridge section had been allowed to settle and seat itself comfortably in its ultimate position on the bridge towers, the precisely located holes in its bottom plates having neatly matched and slipped down over the protruding anchor bolts that had been awaiting their arrival. It was nothing less than a spectacular demonstration of the ingenuity and skills of the bridgemen, and over the months that followed they would erect the other eighteen sections of bridge deck in the same way.[18]

It would still be a long time before that crew of deserving ironworkers at Haverstraw would be heading home. Those big sections of bridge deck they were floating down to Nyack would cover only about a third of the structure's great length. Although other American Bridge crews had by then erected the more straightforward deck framing for the long trestle that connects the bridge to the Nyack shoreline, the most complex job of all still remained to be done: erecting the big trusses that would span the main shipping channel and the anchor sections that would flank it. The deadline for completing the bridge, originally set for late 1954, had been rescheduled for late 1955, less than a year away. Much of the steel that was still required would be prepared in Haverstraw, but placing and securing it in the yawning gap that remained to be filled, hundreds of feet above the Hudson, was a job that would test even the Bridge Company.

This last phase of the job would be every bit as demanding as anything that had come before it. New, even larger falsework sections had to be prepared at Haverstraw and towed downriver, and the main shipping channel was soon filled with barges and scows loaded with thousands of tons of steel.[19] The ironworkers at the bridge site started with the two main towers that flank the channel. Built entirely of steel, they would be the fulcrums

FIGURE 13 The Tappan Zee Bridge, between Nyack and Tarrytown. Photograph courtesy of Paul Scharff Photography (www.paulscharffphotography.com).

on which these giant trusses would be balanced, and once the difficult work of erecting the towers was finished, erecting the trusses themselves could begin. As the steel framing inched its way out in both directions from each tower, the real profile of the bridge structure began to emerge for the first time. By the first of August, the two giant assemblies, almost thirty stories high, were perched atop the towers, their arms reaching out in two directions for support. By the middle of September, it was all over. Connections to the adjacent anchor spans had been secured, and the cantilevered arms had found each other at mid-span, 135 feet above the water.

The ironworkers could go home, but there were still loose ends to be tied before the bridge would be complete. Although paving, painting, and the installation of lighting and other support systems had been started earlier in the year, it would be a race to the finish to be ready for opening ceremonies on the fifteenth of December 1955. On that wintry Thursday politicians, including Governor Averell Harriman, traveled through an upstate snowstorm to tell a big crowd in Tarrytown how happy they were to see the bridge finally opened to traffic. The celebrating was loud and exuberant, the weather cold, and the speeches long and more or less conventional. One speaker identified Bertram Tallamy as "father of the bridge," even though that capable civil engineer and former superintendent of the Department of Public Works had been named chairman of the Thruway Authority in 1950, after most of the spadework had been done. Emil Praeger was neither present nor publicly mentioned, nor was former governor Thomas E. Dewey, who had started it all. (Dewey reportedly wired his congratulations to Governor Harriman.) There was no formal attention paid to any of the ironworkers who had done such extraordinary work, either. A small contingent of protesters picketed the ceremonies and was gently acknowledged by Governor Harriman, but the leading citizens of the hamlets that had been adversely affected by the bridge's location refused to show up at all.[20]

Because there had been no formal agreement about what to call the bridge, it was never referred to by name at the opening ceremony, although the press took up the name that had by then come into fairly common usage: Tappan Zee Bridge. In February 1956 the New York State Legislature, rejecting the Governor George Clinton Bridge, the Alexander Hamilton Bridge, and a few other suggestions, made Tappan Zee Bridge the official name.[21]

CHAPTER 12

The Kingston-Rhinecliff Bridge

BY THE TIME VEHICLES FIRST BEGAN TO CROSS the Tappan Zee Bridge, in the winter of 1955, construction of another Hudson River bridge, seventy-five miles farther north, was well along. It would connect Kingston, New York State's first capital, with the northwesterly corner of New York's Dutchess County.

It's not surprising that the Kingston-Rhinecliff was one of the last of the river's bridges to be built. It didn't have the advantage of lying along the route of the New York State Thruway, as the Tappan Zee did, and in fact, it didn't really have much in the way of other major highway connections to justify it either. There was a modest amount of industry in the area, but during the difficult days of the Depression that certainly hadn't been enough to convince the legislature that money should be spent there for a bridge. Depression conditions notwithstanding, the growing use of automobiles during the 1930s had increased the popularity of the Catskill Mountain resort area, which was easily reached through Kingston and Rhinecliff, but wartime restrictions on the use of gasoline quickly reduced that travel. Once the war got underway, few if any projects not essential to its prosecution got much attention, and the idea of a bridge between Kingston and Rhinecliff languished.

Of course, people in and near Kingston and Rhinecliff had been getting across the river by ferryboat for almost 250 years by then, and they had for the most part accepted its weaknesses along with its strengths. The evolution of that system, which was older than most, had pretty much followed the historical pattern of many of the Hudson's other ferries, starting with a periauger, a small vessel that relied on a couple of hollowed-out logs for a hull and could be driven by either sail or oars. After a few years, horses walking on a treadmill turned paddlewheels. Some time around 1845, the owners of the local ferry company replaced their still primitive boats with a steam-driven vessel that carried many more passengers and wagons and substantially improved the ferry's reliability by providing virtual independence of wind and tide. Those early ferryboats took passengers and freight back and forth between Rhinebeck on the east side of the river and Kingston Point and Rondout on the west side.

Within about a dozen years of the introduction of steam power for the ferry, the construction of the Hudson River Railroad along the east bank of the river strengthened the ferries' passenger base a bit, and issues like the location of their terminals and their scheduling began to be resolved. In 1851, just after the railroad had been completed all the way from New York to Albany, a Hudson River Railroad director named Charles Russell acquired a big tract of east bank riverfront land called Kipsbergen, near the village of Rhinebeck, subdivided it, and began an effort to sell off construction sites. He bought the ferry company, too, moved its dock to Kipsbergen, induced his fellow directors at the railroad to establish a station conveniently close to where the ferry landed, and named his emerging subdivision "Rhinecliff."

Over the ninety or so years that elapsed between those events and the time when serious talk about a bridge to Kingston began to be heard, Rhinecliff and its neighbors in Rhinebeck prospered independently, but Rhinecliff's dominance as the transportation hub didn't generate much enthusiasm in the surrounding communities. The Hudson River Railroad waited years before changing its timetable to show that its trains would stop at Rhinecliff, not Rhinebeck, and the New York and Albany Day Line never changed its schedule.[1]

By the 1930s, the popularity of what the public had come to know simply as the Kingston-Rhinecliff Ferry was wearing a little thin. Its customers had long accepted its regular winter shutdowns and limitations on its hours of operation during the rest of the year as natural and unavoidable, and they were for the most part tolerant of the sometimes prolonged delays imposed by bad weather and mechanical failure. But between about 1930 and 1940, when even under the severely depressed conditions of the period the number of motor vehicle registrations in the country had increased by about six million, many people in the area had acquired cars of their own and had begun to think about how nice it would be to drive across the river at any time of day or night and in any kind of weather. It was hard to ignore the Mid-Hudson Bridge, about fifteen miles south, or the Rip Van Winkle Bridge, about twenty miles north, both of which provided those conveniences, and the possibility of a bridge instead of a ferry between Kingston and Rhinebeck began to attract more public interest.

By late 1942, the ferry's circumstances had deteriorated so badly that what had been passive interest in the idea of a bridge had begun to turn into an aggressive campaign to build one. The ferry's volume of automobile traffic, decimated by gas rationing, had fallen by more than half, and in December river ice had shut it down for the season. In March 1943 the ferry company announced it wasn't ever going to resume service, and in November it sold its boat to Harris County, Texas. In 1944 it elected to dissolve, leaving Rhinebeck and Kingston high and dry.[2]

When that unpopular decision was made, Arthur H. Wickes was a fifty-seven-year-old laundryman in Kingston who had by then served for seventeen years as a senator in the New York State Legislature and was destined to serve for another eleven. Despite a predilection for rarely saying or writing anything in ten words that could be said or written in twenty, Wickes was a committed and effective public servant, sensitively attuned to the needs of his constituents, and he took on both the ferry problem and the bridge campaign.

Almost as soon as the ferry company announced its closing, Wickes introduced a bill to authorize the New York State Bridge Authority to take over its operations. By spring 1946 the authority had acquired a ferryboat capable of carrying thirty cars and 305 passengers, named it the *George Clinton* (after the Kingston native who had been New York's first governor), recruited a staff (mainly from the old ferry company), and restored service between Rhinecliff and Rondout, a hamlet just south of Kingston. The restoration of service, after almost three years without it, was welcomed by the populations of and near Rhinecliff and Kingston and by other potential passengers, but most people were well aware that the ferry was still at the mercy of ice, bad weather, and mechanical failure, and they had no illusions about its reliability or its permanence. It was a welcome stopgap, but the campaign for a bridge continued.

When it came to getting the design and construction of a bridge authorized, Senator Wickes would have his work cut out for him. Between 1944 and 1946 he and State Assemblyman Ernest Hatfield were able to get two bills passed authorizing the New York State Bridge Authority to start the project, but both bills were vetoed by the governor, who felt that some problems still needed to be sorted out before he could give his approval. There was some merit in what he said. Not the least of them was a commitment that the authority had made years earlier to the bondholders of the neighboring Bear Mountain, Mid-Hudson, and Rip Van Winkle bridges, guaranteeing that no more bridges would ever be built along the Hudson between Albany and New York. And that wasn't all. There was also the matter of $7.5 million the Bridge Authority still owed the State Treasury for its share of the original cost of building the Mid-Hudson Bridge.

Between some difficult and effective negotiation and some creative legal work, Senator Wickes would by late 1946 manage to solve both those nasty problems. Early in 1947 the governor signed a bill directing the Bridge Authority to start design work for what would (out of respect for the old ferry and regardless of the actual location of the new bridge) later be named the Kingston-Rhinecliff Bridge, and he had arranged to find $50,000 with which to get the design underway. The authority was well established by then, successful and brimming with confidence, and it moved quickly to engage engineer David B. Steinman to do the necessary preliminary engineering work, including feasibility studies, surveys, and a conceptual design.[3]

Steinman, sixty years old when he took on the Kingston-Rhinecliff commission, was the eminent founding principal of a large and successful firm that had designed more than three hundred bridges, many of them landmark structures, during the twenty-six years that preceded the Kingston contract. A New York City native, he had been educated at the City College of New York and at Columbia University, where his 1910 doctoral thesis on the design of suspension bridges and cantilevers would later be commercially published and marketed as a text. His first few years after leaving Columbia were spent teaching engineering at the University of Idaho, where he was half the school's two-person civil engineering faculty. After four busy and broadening years there, Steinman gave up riding to class on a horse and enduring winters at twenty degrees below zero to accept an invitation to work as one of two special assistants to the venerable bridgebuilder Gustav Lindenthal back in New York. There, working on Lindenthal's Hell Gate and Sciotoville bridges, he would get his first experience in designing large and important bridges. In the heady environment of such projects, Steinman would during his years with Lindenthal meet and first cross swords with the brilliant Othmar Ammann, the second of Lindenthal's special assistants and the future chief engineer of the George Washington

Bridge. The profoundly contentious relationship that developed between the two young engineers would never be eased.[4]

Steinman's career since the Lindenthal years had been immensely productive and at times spectacular. Early along, he and his occasional partner, Holton Robinson, had designed the approximately 1,100-foot suspension bridge that connects the Brazilian state of Santa Catarina with its island capital of Florianapolis, a project that was complicated by the irresponsible performance of an unreliable foundation contractor and the perfidy of a dishonest financial manager. Steinman's use of eyebar construction in the suspension system at Florianaoplis would ironically provide a useful model years later when Ammann considered it as an option for the George Washington. At California's Carquinez Straits Bridge, where Glenn Woodruff (who would later design the Rip Van Winkle Bridge) joined the firm, Steinman did his first important work in designing for severe earthquake conditions. Many other bridges in widely separated locations followed, most of them suspension bridges, and by the late 1940s Steinman was pleased to be able to say that he had built bridges on every continent except Africa. When the Kingston project came long, he felt as strongly as ever that the suspension form should be used wherever it could be justified.[5]

In October 1948 Steinman presented his report for the Kingston-Rhinecliff Bridge to the New York State Bridge Authority. To no one's surprise, he and his staff had found the bridge to be entirely feasible, with ample economic justification that was well supported by surveys, ferry records, and the like. Equally unsurprising was his recommendation that the crossing be a suspension bridge, and he had developed a preliminary design showing a graceful structure that delighted almost everyone who saw it. Its total length was to be 6,960 feet, including its approaches, with 3,970 feet of structure between the anchorages. The suspended span of 1,700 feet was to be as long as the corresponding span of the Firth of Forth Bridge in Scotland, and only six such spans in North America were longer. The bridge's 22-foot-deep stiffening trusses were to be positioned entirely below its deck, to preserve the traveler's view of the valley.

A number of routes were proposed, each not very different from the others, but the one recommended by Steinman's report would run directly through the villages of Rhinebeck, at the east end, and Kingston, at the west end. The report included an estimated cost of $14 million, a figure that seemed high to some who heard it but that didn't seem to shock most of them or portend rejection.[6]

It seemed likely that formal approval would be given soon, but more than a year after the report's submission, the chairman of the Bridge Authority was still publicly acknowledging and defending a continuing delay in its approval, explaining to the local chapter of the Rotary Club that although the authority liked the route proposed by Steinman, the state's Department of Public Works was urging a more northerly route to avoid passing through the village of Kingston. It was suggested that such a northerly route might encourage a connection between the New York State Thruway, which was under construction west of the river, and the Taconic Parkway, east of the river, but that argument seemed tenuous to some. Both Steinman's route and a more northerly one seemed about equally capable of providing such a connection.[7]

Not much attention was given to the delay in the local press, and it would be almost another year before the impasse would be resolved. Finally, in August 1950, the Bridge Authority announced that the route proposed by Steinman had been rejected in favor of one almost four miles north of it that would avoid Kingston entirely. That decision would radically affect the bridge's design and its schedule.[8]

Although the reasoning behind the change of route was never fully explained at the time, it seems clearer now. The decision was made during a period in the late 1940s when the real implications of the coming age of electronic computing were just beginning to be seen and understood outside the scientific and academic communities, and nowhere were they clearer than in the Hudson Valley. Until then, International Business Machines (IBM), the valley's largest employer, had been a relatively small player in the electronic computing field, having pretty much concentrated its engineering capability on its military commitments during the war and having since the war's end returned to its traditional, commercial product line. But in 1945 the University of Pennsylvania had unveiled a 30-ton electronic computing machine that could add or subtract 5,000 times in one second, and young Tom Watson of IBM had gone down to Philadelphia with some of the company's most sophisticated scientists to have a look at it. There had already been some talk among IBM's managers about becoming more vigorously involved in electronic computing, but seeing what had been done at Penn probably added a certain urgency to that thinking. By 1946 one of the company's most capable researchers had been placed in charge of an expanding product development laboratory at IBM's sprawling plant in Poughkeepsie, expressly to intensify its focus on electronics.

During what remained of the 1940s, the pace of development was significant, and by about 1950 what was emerging had begun to portend a shift in the company's direction that was destined to bring it to a position of dominance in the manufacture of electronic computers.

Poughkeepsie, only about fifteen miles south of Kingston, became an early center of production, and there was a growing likelihood that the change in direction there would soon crowd out the plant's extensive commitment to the production of electric typewriters. Anyone on the inside of things at Poughkeepsie probably recognized that it wouldn't be long before IBM would need a lot more space than it had (or could build) in Poughkeepsie, and it's not unlikely that a good deal of thought was already being given to building in nearby Kingston.[9]

The secrecy that attaches to that kind of planning is guarded carefully in companies like IBM, often for good reasons: It could be useful information for competitors, it could be disturbing to some employees, and it could inflate the cost of land that might later have to be acquired. There was certainly very little publicity given to the idea of a new IBM plant in Kingston, and nothing was mentioned in the Kingston press. It's doubtful that David Steinman or any of his engineers ever gave the possibility a thought. If they had, they would never have located their proposed bridge where it would bring traffic into the heart of the small village of Kingston, just when a whole new population was about to be added to it. As things turned out, IBM would acquire its first two hundred acres in Kingston in February 1954, start construction of what would become about 500,000 square feet of plant space in May, and begin operations in Kingston in 1955. By the time the Kingston-Rhinecliff Bridge was completed, there would be almost five thousand people working in the new plant, and many of them were likely to be traveling across the new bridge twice a day.[10]

The decision to move the bridge route almost four miles north saved Kingston from a traffic nightmare, but it invalidated much of what Steinman and his staff had done. The justification for building any bridge at all had certainly been strengthened, but the geologic and topographic information that had provided the basis for Steinman's design was no longer relevant at the new location. A contract to design the new bridge was soon awarded to the Steinman firm, but David Steinman's role in the work would be less central than it had been. Although the river was only a little wider at the new site, the existence of dual shipping channels separated by a wide shoal area and the geology of the riverbed there discouraged construction of a suspension bridge. The first task of the designers would be to identify the best alternative. The abandonment of the suspension scheme had to be a serious disappointment to a man for whom suspension bridges had become central to his career, but it's doubtful that it had any bearing on Steinman's decision to reduce his role in the Kingston job. Other events were more compelling.

In fact, Steinman had by then been engaged for many years in a determined but unsuccessful effort to secure the commission to design a bridge to connect the lower and upper peninsulas of Michigan, across the Straits of Mackinac. That bridge promised to be one of the most spectacular suspension structures in the world, with a clear span longer than that of Ammann's George Washington Bridge, at that time the world's longest. Steinman had done about everything he could to secure the commission, but during the 1930s the contract was awarded to the firm of Modjeski and Masters, which had arranged for the eminent Leon Moisseiff to head up its design team. When Moisseiff's Tacoma Narrows Bridge suffered a catastrophic failure a few years later, the Mackinac authorities decided to reconsider their relationship with Modjeski and Masters, opening up the possibility that Steinman might have another chance. Then came World War II and the suspension of almost all nonmilitary work, but soon after the war ended, interest in building the great span in northern Michigan had been revived, and the Michigan people took another look at David Steinman. By the time the New York State Bridge Authority got around to authorizing Steinman to proceed with a revised design for the rerouted Kingston-Rhinecliff, he had been named to design the Mackinac. The big bridge across the Straits of Mackinac would be five miles long, would cost about $100 million to build, and would have a suspended span of 3,800 feet, a world record that would survive until the last years of the twentieth century.[11]

By 1950 Steinman's firm was designing an average of about twenty bridges per year, and it had no problem producing a couple of well-qualified engineers with many years in its employ to step in at Kingston. Walter Joyce, who had begun his career with the New York State Department of Public Works, took charge. He had started with Steinman almost twenty-seven years earlier and had become his principal associate. Carl Gronquist, who had joined the firm only a few years later than Joyce, became chief of design for the Kingston bridge. The original studies were dusted off, borings were started along the new route, and by early 1951 design work for the new bridge was underway. It would be a long process. Between the demands of having to rework all the calculations for the new site and the need to absorb delays that are not uncommon in large, public-sector work, it would be almost three years before a completed set of construction documents would be ready for presentation to contractors for bidding.

The concept that emerged in the spring of 1954 was a continuous deck truss design that included arched bottom chords for the very deep trusses that bridged the river's two shipping channels, and continuous, shallower deck trusses for the rest of the river crossing. The shorter spans along the viaducts that reached out across adjacent land areas to connect the river spans to the local road networks were designed with deep plate girder construction. The river at the new location was almost a mile wide and contained two approximately 800-foot-wide shipping channels, one in its easterly section and the other in its westerly section, separated by an approximately 1,700-foot-wide shoal area in the middle of the stream. The new design maintained the 800-foot-wide clearance mandated by the Corps of Engineers for navigation in each shipping channel, and it provided the required 135-foot vertical clearance above mean high water as well. In the approved scheme, which had not been the least expensive option, all the truss work would again be kept below the bridge deck to preserve the unique upstream and downstream views that are especially prized in that section of the river. The bridge road itself would be 36 feet wide to allow for three lanes of travel and a couple of two-foot-wide emergency walks. The bridge's overall length, including its approaches, was about a mile and a half.[12]

All the construction, except for a few minor components like an administration building, was concentrated in two large contracts, one for foundations and related substructure and the other for the superstructure. The successful bidders would be no strangers to bridge work on the Hudson. The low bidder for the foundation contract was Merritt-Chapman & Scott, which was just finishing up its big contract on the Tappan Zee Bridge, and the low bidder for the superstructure was Harris Structural Steel Company, which had twenty years earlier built the steel superstructure of the Rip Van Winkle Bridge.

The timing of the foundation bidding was fortuitous. A brief recession had intensified competition for construction contracts during 1953 and 1954, and eight firms submitted bids for the Kingston-Rhinecliff job, estimated by the engineers to cost about $5.5 million. The Merritt-Chapman price of $4.5 million surprised just about everyone. It was about $800,000 lower than the second-lowest bid, and almost exactly half the price proposed by the highest bidder. Only a few months later, Merritt-Chapman would be awarded an approximately $26 million contract to build the foundations and other substructure for Steinman's mammoth bridge across the Straits of Mackinac. It's doubtful that the company would have been so sharply competitive in the Kingston bidding if it had known it would be doing the big Mackinac job.[13]

Bidding for the superstructure contract doesn't appear to have been intensified by the recession. Although it was twice the size of the foundations contract, it wasn't quite big enough to catch the eye of any of the fiercely competitive national steel companies, but it was too big for most of the local fabricators. Only three firms submitted proposals, and the narrow range of their prices was seen as evidence that Harris Steel's low bid of about $10 million should be the basis of an award.[14]

Like most marine construction, the job that lay ahead for Merritt-Chapman & Scott was tough work from start to finish, and it would be about eighteen months before the substructure was far enough along to allow Harris to start erecting its structural steel. The bulk of the foundation work involved building the bridge's thirty supporting piers and the concrete towers above them, tasks that required approximately equal amounts of demanding and hazardous work below the water's surface and above it. Bedrock for supporting the piers in the river was exceptionally deeply buried, some of it down as much as 150 feet below the water's surface. For the four piers that were to carry the heaviest loads (adjacent to the wide shipping channels) the contractor successfully used the dredged, open-caisson method that had been used (with mixed results) at the Mid-Hudson Bridge. For the three river piers in the shoal area, Merritt-Chapman was required to build big concrete pile caps and to support them on cast-in-place concrete piles reinforced with structural steel columns and driven to rock. At most of the viaduct piers, where shorter spans imposed lighter loads, acceptable bearing was found at shallower depths; most of those piers were supported by concrete piles driven to refusal, and the rest were founded directly on rock without piling.

As the substructure work inched its way to completion, around the middle of 1955, Merritt-Chapman still had plenty of other work to do. The bridge deck itself would rely for support on towering reinforced concrete bents, and the 4 percent grade of the bridge meant that some of them would reach heights as great as 194 feet. That was all part of the foundation contract. By fall, with winter weather threatening and Harris due to start in the spring, everything possible was being done to get some of those bents ready, and it wouldn't be easy.

In the early months of 1956 Harris was marshaling the men and material it would need to erect the steel superstructure, but earlier design and construction delays had taken their toll. The bridge had been promised for November 1956 (after several postponements), and with about 15,000 tons of steel to be erected before paving, painting, lighting, and other work could be done, meeting that commitment looked virtually impossible. To make matters worse, winter conditions had left an unusual amount of ice in the river, limiting Harris's use of barges to bring material to the site, and a wet spring and muddy conditions ashore made deliveries of steel across the land areas equally difficult. At that point, Harris might even have begun to feel some relief that Merritt-Chapman wasn't quite ready for steel.

But by the end of winter, the steel work was finally getting started. Harris had made a strategic decision to use two competing crews to erect it, one on the easterly section of the bridge and the other on the westerly section. The two crews started their work at the east and west viaducts, where the shorter spans had made it possible for the engineers to use plate girders (some as long as 150 feet and as deep as 11 feet) instead of deep trusses. By the time that work was well advanced, the weather had started to improve. The plan for erecting the big steel trusses was to leave the channel spans for last, as the cantilever method was to be used there as a way to avoid the use of falsework in the channels. But there was no constraint about using falsework for temporary support in the other spans, so the two ironworker crews made their way east and west, starting at the channel spans. First they erected temporary pile-supported bents and then moved out across them with traveler cranes to assemble and erect the 50-foot-deep trusses that would carry the bridge deck. By the middle of October, that work having been done, the cantilevers of the channel spans were erected and jacked into their final positions, closing the 800-foot channel gaps that would finish the superstructure.[15]

Completing that big steel job in what was effectively about six months was exceptionally good work and testimony to the skill of the ironworkers who did it. Bridge ironworkers have always been a rugged lot, of course, and in the case of the Kingston-Rhinecliff Bridge their effectiveness was enhanced by the presence among them of a few Mohawk Indians, members of the Kahnawake community in Canada. It was no accident that there were no more than a few Kahwanake working on the Kingston-Rhinecliff. Since 1907 it had been a tradition in the community that they would never again concentrate a large number of their people in the construction of a single bridge.

The Kahwanake had first come to bridge building in 1886, when the Dominion Bridge Company of Canada hired some of them as ordinary laborers on a bridge the company was building across the St. Lawrence. Noticing the extraordinary talent the Indians showed for walking the high steel without fear or hesitation, the company's managers decided to train a few of them to do ironwork, and the results exceeded their most optimistic expectations. For reasons that have never been fully understood, the Kahwanake proved to be entirely fearless in performing complex tasks on narrow steel members at great heights, and they were fast learners, too. Within a short time they proved to be as skilled and as highly valued as Dominion's most experienced ironworkers.

More Kahwanake entered the trade after 1886, and in the summer of 1907 many of them were working on the construction of an even larger steel bridge across the St. Lawrence, the great cantilever near Quebec. On the twenty-ninth of August the unfinished structure collapsed, killing seventy-five ironworkers, thirty-five of them Kahwanake. It was a tragedy that decimated the small community's male population and left numerous widows and orphans. Although there was a clear determination by the survivors to continue working in the high steel, tribal elders decided that any concentration of members of the community on a single bridge would be avoided in the future.

The Kahwanake stayed clear of almost all steel work for about a dozen years after the Quebec disaster, but during the 1920s many of them began working in the high steel of the skyscrapers that were proliferating in New York and other large American cities. Not surprisingly, they distinguished themselves there, too. Although they never returned to bridge work in their former large numbers, some Kahwanake are said to have worked on a few of the New York City bridges, including the Hell Gate, the George Washington, and the Triborough, but there's no evidence that they figured prominently in the construction of bridges along the central or upper Hudson before the Kingston-Rhinecliff came along.[16]

The completion of the steel on the Kingston-Rhinecliff in October 1956 didn't mean the bridge would open that year. There were just too many other things left to finish, and some of them (like the concrete paving of the deck) would be extremely difficult to do during the winter months. But a great effort was made to finish as quickly as possible, and by the end of January 1957, aided by an abundance of canvas covering and kerosene-fueled heaters, the work was brought to a condition that was seen as substantial completion. The level of public impatience having risen precipitously, an informal opening was held on the second of February.

The crowd at the opening was addressed by Governor Averell Harriman, who delivered his remarks during a stop on his way to a ski vacation, and he added to the informality of the proceedings by arriving in ski clothes. The bridge, which had cost almost $18 million, was formally but imprecisely named the Kingston-Rhinecliff for the first time. The legislature had legalized that name a few months earlier as a respectful nod to the old ferry the bridge had displaced, but in fact the bridge's eastern terminus was Rhinebeck.

FIGURE 14 The Kingston-Rhinecliff Bridge. Photograph courtesy of the New York State Bridge Authority.

The slightly premature opening was attended by many of the politicians who had figured in the bridge's history. Although there's no evidence that David Steinman was present or acknowledged, William Joyce, who had shepherded the project for the Steinman firm from its beginning, was given a prominent position in the two-hundred-car motorcade that was first to cross. The newly placed concrete deck was covered by a dusting of snow, the steel superstructure hadn't yet been painted, bridge lighting was still to be installed, temporary wood curbing substituted for concrete, and temporary toll booths had been erected for the occasion. But most people were glad to be able to drive across a bridge they had awaited for thirteen years.[17]

Three months later, on the eleventh of May 1957, when things were in much better shape, a formal dedication took place. Like most second openings, it was poorly attended, and the meager festivities were dampened by a continuing rain. But the event provided a symbolic date on which the bridge could be said to have been completed.[18]

CHAPTER 13

The Newburgh-Beacon Bridges

WITH ONE GLARING EXCEPTION, completion of the Kingston-Rhinecliff Bridge in 1957 meant that travelers could cross the Hudson at or close to almost every one of the valley's busiest towns and cities. The exception was an approximately thirty-mile stretch that still separated the Mid-Hudson Bridge from the Bear Mountain Bridge, along which the bustling Hudson Valley town of Newburgh was about in the middle on the west side of the river, and the quieter, more residential town of Beacon was opposite it on the east side. There was no bridge at Newburgh.

Just why a bridge connecting the city of Newburgh with its easterly neighbors in Beacon had never been built, or why it would be the last of the bridges to cross the river during the twentieth century, isn't easy to understand. It certainly wasn't because Newburgh wasn't a good candidate for a bridge. It had been a thriving Hudson River town since the Revolutionary War, when George Washington made it his base of operations for a while. Its considerable concentration of light industrybegan to develop in the nineteenth century, when brick manufacturing plants and mills for processing wool, cotton, paper, and a variety of other materials began to appear and flourish, providing a combination of industry and agriculture that led to its prosperity as a thriving river town. It is said to have been the first city in the country to become electrified, and during its best years its retail commerce was the envy of it neighbors. Never as big as Poughkeepsie or Albany, of course, it was about the same size as other upstate bridge towns like Hudson and Kingston.

And it wasn't because the idea of a bridge between Newburgh and Beacon hadn't been given plenty of thought. As early as 1925, General Milton

Davis, a locally popular, decorated veteran of several wars and the superintendent of the New York Military Academy in Peekskill, proposed such a bridge in a public meeting. He later mounted a modest campaign for it, but nothing came of his proposal.[1]

Opposition from competing towns along the river was clearly a factor at different times, but it was hardly decisive. When the original Poughkeepsie Railroad Bridge was being planned back in the nineteenth century, Newburgh boosters unsuccessfully campaigned for routing the railroad through their town instead, but Poughkeepsie won the day. Later, when the idea of a bridge at Bear Mountain was proposed in the early 1920s, Newburgh made an attempt to raise capital for a bridge of its own, but again it was unsuccessful. More recently, when legislation for the Kingston-Rhinecliff Bridge was enacted, the legislature included a provision that barred the construction of any other bridge across the Hudson until the bridge at Kingston was completed. That was a nearly lethal blow to any near-term plans for building at Newburgh, which appears to have been a candidate whenever a nearby bridge was being considered but which had never managed to win the prize.[2]

It's possible that the real reason for all this hard luck might have been the genuine affection that the populations of Newburgh and Beacon had for so many years felt for their ferry. Established by royal charter to Alexander Coldenham in 1743, the ferry was in 1935 said by the *Newburgh News* to be the oldest ferry in America. Maturing from oars to horses to sails and ultimately to steam power, it was still providing acceptable transportation between Newburgh and Beacon when talk about a bridge first began to be heard in the 1920s. Fog, ice, darkness, mechanical failure, and occasional fires notwithstanding, the ferry had become more than a mere convenience to people in and around the two towns and had in some cases become central to their lives. It wasn't unusual for visitors from the east side of the river to come across to meet friends and family in the ferry's Victorian-style terminal in Newburgh, spend time browsing in the upscale shops that lined the 132-foot-wide, brick-paved boulevard that led into the commercial center of Newburgh, and then to take the ferry back to Beacon before dark. Merchants, professionals, and others had relied on the ferry for years, and most of them were prepared to continue patronizing it until they were convinced that a better option was possible. There's no evidence that any of these loyal ferry travelers ever participated in an organized resistance to building a bridge, but it's not unreasonable to think that their affection for the ferry (and their corresponding disinterest in a bridge) was something that discouraged support from political figures whose power and influence would be required if a bridge were ever to be built.[3]

Proposals for such a bridge continued to surface from time to time during the thirties and early forties, until the war came along and refocused everyone's attention, and there's little evidence that the concept ever developed any real traction before the end of the war. That was when a surge of returning veterans would infuse local thinking with their own ideas and their eagerness to make up for lost time. In 1945 the Orange County Chamber of Commerce named a committee to develop a serious proposal for building the Newburgh-Beacon Bridge, and its report marked the beginning of a campaign that would gather increasing momentum over the next ten years and would ultimately succeed.[4]

Early in the 1950s, there was some evidence that passenger loyalty to the ferry was beginning to wane. A new crop of travelers wasn't as tolerant of delays and cancellations as their forebears had been, and when one of the ferries ran aground and a second lost its rudder in a rescue attempt, the bridge campaign began to win over a few former ferry loyalists. Some of them had by then started using the Mid-Hudson Bridge and the Bear Mountain Bridge, each less than half an hour's drive from Newburgh, and some were finding other strategies. The number of commuters who drove their own cars had increased substantially, and the old ferries simply didn't have the capacity to take many of them aboard in any single trip. Once automobile waiting lines became objectionably long, patronage declined, and the ferry company's once substantial financial base began to weaken.

At about the same time, plans for the New York State Thruway were being developed, and there was some optimism among the bridge advocates that the new highway might cross the Hudson at Newburgh, generating substantial traffic volume and more than enough justification for a bridge, but a decision in favor of crossing at Tarrytown, instead, dashed that hope. Undaunted, Lee Mailler of Cornwall, who was the state assembly's majority leader and a passionate advocate for the bridge, successfully sponsored a bill to provide $50,000 for test borings. It was the first positive evidence that there might really be a bridge after all.

The borings were completed during 1951, and by early 1952 engineers of the New York State Department of Public Works (predecessor of the Department of Transportation) had completed preliminary engineering studies for a two-lane bridge they estimated could be built for a little less than $18 million.[5] Their cost estimate was pretty good, and although their design wouldn't survive, what they had produced proved to be a good start. In fact, the Public Works report found its way to Governor Thomas E. Dewey's desk, and by April 1953 the legislature had passed (and the governor had signed) a bill authorizing construction of the Newburgh-Beacon Bridge.[6]

Time became an increasingly critical factor. The level of ferry service was worsening, and there was widespread (amply justified) uncertainty about whether the operation would survive until a new bridge could be put in place. For one thing, the prohibition against building any more bridges until the Kingston-Rhinecliff was completed was still in place, and work on that bridge was just getting underway and expected to take at least two more years to complete. Complicating matters, the state's authorization to build the Newburgh-Beacon Bridge hadn't come with any money. It was simply a bill that authorized the Bridge Authority to build the bridge, leaving it to the authority to sell bonds to pay for it. That was something that most people knew wasn't going to be done easily or quickly. With such obstacles, it was clear that a bridge between Newburgh and Beacon was still a number of years off, probably more years than the ferry could be expected to survive on its own. There were still hundreds of people in Newburgh and Beacon who relied on it to get them to and from their jobs, so the ferry's survival and planning for a new bridge were a couple of issues that would dominate the agenda of political leaders in Newburgh and Beacon for much of the mid-1950s.

Not everything was going badly. Now that construction of the bridge at Kingston had become a reality, political leaders there apparently felt confident enough to join forces with some of their colleagues in the legislature (including a few from Newburgh) to repeal the troublesome prohibition against building bridges at Kingston and Newburgh at the same time. Soon afterward, the Bridge Authority secured the legislature's authorization to increase its debt limit to a level believed high enough to allow it to fund bridges at both those locations simultaneously.

It was beginning to look like construction might get underway fairly soon after all, but nothing's perfect. Before 1954 ended, a second look at the Bridge Authority's calculations suggested that its projected revenues were not going to be enough to support the higher debt that would have to be incurred if they were going to build the two bridges at the same time, and the Newburgh bridge advocates were obliged to seek another approach.

But the Newburgh cause didn't lose everything when the authority changed course. One small by-product of the authority's new bond issue was a first installment in what would become a $1.2 million fund earmarked specifically for engineering and related work for a bridge at Newburgh. Those funds paved the way in 1955 for the Bridge Authority to engage the Harrisburg-based, nationally prominent engineering firm of Modjeski and Masters to design the Newburgh-Beacon Bridge. The Modjeski firm was no stranger to the Hudson Valley, having designed the Mid-Hudson Bridge at Poughkeepsie and having provided consulting services on at least one other Hudson River span as well.

Modjeski and Masters had in 1893 been established by the eminent Ralph Modjeski, who had retired in 1936 and died in 1940, after his firm had designed many of the country's most important bridges. He had been succeeded by his partner, Frank Masters, who was himself an unusually vigorous seventy-three years old in 1955. Masters had started his engineering education at Cornell University around the turn of the century but had dropped out for financial reasons, and by 1924, when he and Modjeski became partners, he had in the tradition of the times acquired and refined his expertise in civil engineering through employment and collegial relationships with civil engineers. By 1955, when the Newburgh-Beacon work was started, the firm had under his direction designed hundreds of bridges, and Masters himself had achieved a reputation as distinguished as Modjeski's, if not quite as widely known.[7]

The engineers lost no time getting started on surveys, studies, preliminary designs, and cost estimates for the Newburgh-Beacon, although there was still no money for building it. Those in charge for the state were openly optimistic about the financing, and they authorized Modjeski and Masters to proceed with the work as if all the money were in hand.

While the design was taking shape, time ran out for the ferry. Around the middle of 1956, more than two hundred years after the ferry's eighteenth-century beginning, its owners let it be known that they would soon be shutting down. Before the year ended, the protests and pleas of outraged ferry customers, some of them hospital workers, policemen, teachers, and others whose ability to cross the river was seen as vital for maintaining community services, had persuaded the governor to direct the New York State Bridge Authority to step in and take over the ferry company.[8]

Lots of what was going on in 1956 would affect the bridge, and not all of it was happening in Newburgh, in Beacon, or even in Modjeski's office in

Harrisburg. In Washington, DC, the Bureau of Public Roads was laying plans for a major extension of its Interstate Highway System, and Congress was putting the finishing touches on an expanded Federal Highway Act. In Albany, Averell Harriman had succeeded Thomas E. Dewey as governor, and his administration was doing everything it could to ensure that New York State would get its share of whatever the feds were planning.

What was going on in Washington seemed made to order for the Newburgh-Beacon Bridge. Dwight Eisenhower had been elected for a second term, in 1956, and he had brought to the presidency a serious commitment to an interstate highway program. Congress had started him off on the right foot during his first term with the Federal-Aid Highway Act of 1952, which substantially increased federal aid for 1954 and 1955 by promising to match state funds, dollar for dollar, to pay for qualifying interstate highway construction. In 1954, the government had raised its share to about 60 percent, depending on local circumstances, and by 1956 all the earlier bills had been consolidated and broadened into what would become the landmark Federal-Aid Highway Act of 1956. In that legislation, the feds were talking about grants as high as 90 percent of the cost of qualifying projects, and with such a source of funding emerging, it's not surprising that a waiting line was already forming. Long lines or not, something that made that news from Washington especially good for Newburgh was the contemporaneous report that a new four-lane interstate highway, designated Interstate 84, was going to cross the Hudson between Newburgh and Beacon.[9]

Nothing definite had been committed yet, and there were complications. One important one was the federal government's continued resistance to the collection of tolls on interstate roads, although the language of the new Highway Act suggested a little more flexibility on that issue than had been seen in earlier legislation. It implied a willingness to make exceptions to its toll constraints wherever there was a reliable prospect that the tolls would be abolished once the debt they were intended to amortize had been repaid. Unfortunately for New York, its own administration apparently wanted to preserve its right to continue collecting its tolls longer than that, perhaps for application to other projects, and that was a factor that appeared to be limiting the state's prospects for funding.

In June 1956, in an atmosphere that was optimistic but still uncertain as to details, Modjeski and Masters submitted a report to the state that included preliminary designs that could be used whether New York received federal funds or not. Because the timing of the report had been deferred to allow for passage of the 1956 federal highway bill, its authors were aware of the potential impact of Interstate 84 and were able to reflect its traffic volume in their calculations. The report, in addition to recommending a site and evaluating soil borings, included as its basic scheme a preliminary design for a four-lane cantilever bridge, and an estimate that building it would cost about $35 million, in 1956 dollars, not including anything for property acquisition, engineering fees, or the like. Apropos the engineers' uncertainty as to just how the bridge would be funded, their report also included preliminary designs and cost estimates for three alternative schemes: a six-lane bridge, for about $42 million; a three-lane bridge, for about $28 million; and a two-lane bridge, for about $24 million.

Unlike many preliminary schemes, this one would survive the occasional and sometimes disturbing surprises that can surface during later phases of design development, and the basic design scheme that Modjeski and Masters proposed in 1956 would be remarkably like what would eventually be built, except for the width of its roadway. The substructure would comprise mostly open-dredged concrete caissons (at least one as deep as 165 feet), topped by stone-clad concrete bases to support the steel towers of the superstructure. The superstructure would consist essentially of a deck-type truss system, in which trusses averaging almost 450 feet long would be positioned directly below the level of the bridge deck, except where they crossed the approximately 1,000-foot-span that would ensure an approximately 800-foot-wide navigation channel between the supporting piers. There, they would become through trusses, rising above the deck and reaching toward each other in graceful arcs, ultimately linking with one another through connecting trusses 135 feet above the river.[10]

Governor Harriman liked the four-lane bridge, and he apparently did what he could to negotiate a satisfactory arrangement with the federal government for funding it. But the issue of tolls remained contentious. By late 1957 most attention was focused on just how many years it would take for Washington to turn over to New York the approximately $26 million the state would need to supplement its own funds. A variety of other funding options were considered, including a loan by the state to the Bridge Authority, but the governor didn't like any of them. Finally, in 1958, reacting to a petition signed by 35,000 people, the legislature authorized another million dollars to continue engineering work and land acquisition, while the matter of just how the bridge itself would be paid for remained unresolved.[11]

Nelson Rockefeller, a man with an especially well-honed taste for monumental construction, succeeded Averell Harriman as governor, and not long after taking office in 1959, he embraced the bridge cause as his own. By then a decision for the federal government to fund the Newburgh-Beacon had been made, but a new obstacle had surfaced. An amendment to the Highway Act prohibited the government from advancing or guaranteeing any funds until the revenues from which they would be paid were safely in the government's coffers. Most of the money for the Newburgh-Beacon enterprise was to have come from taxes on motor vehicles and fuel, and the pace of those collections had declined in the late 1950s, delaying the actual disbursement or guaranteeing of highway funding. It was becoming evident that it was still going to be some time before federal funding for the Newburgh-Beacon could be guaranteed, and neither the New York State Department of Public Works, which would be responsible for building the bridge for the Bridge Authority, nor the governor himself was willing to proceed with construction without certainty about the money.[12]

But Nelson Rockefeller was not a man easily diverted from his goals, and he didn't quit. He recognized that building the more modest two-lane version of the bridge, for what Modjeski and Masters were estimating should be about $24 million, was something the state might be able to manage without the help of the federal government, and that work on it could be started almost immediately. The rationale for a four-lane bridge had been based on a prediction that traffic from Interstate 84 would soon be generating a volume of almost nine million vehicles per year, and that was a volume that would produce an intolerable backup for anything narrower than a four-lane bridge. But Rockefeller had an idea that the estimate was probably too conservative, and that the nine million vehicles per year volume was going to be a long time coming, so he turned for advice to John L. Edwards, a Hudson physician who was at that time serving as chairman of the New York State Bridge Authority. Dr. Edwards told the governor just what he wanted to hear. In his view it would be twenty-five years before such a volume would be reached. That was good enough for Rockefeller, and he ordered that the design be revised to provide for a two-lane bridge. Half a loaf was better than none, he reasoned, and New York would go it alone.[13]

That was in 1960. Although a preliminary scheme for a two-lane bridge had been included as an option in the report submitted by Modjeski and Masters in 1956, it hadn't been fully developed, and there was still plenty of work to be done to convert it to a set of construction documents. The engineers went right to work, and by summer they were beginning to produce drawings that could be submitted to contractors for bidding.

The governor's office had its work cut out for it during those months. The Bridge Authority had by then become a large and powerful agency, with jurisdiction over five Hudson River bridges, and the legislature had recently increased its bonding capability to $70 million. Rockefeller had reasonably anticipated that with such an increase, the authority would raise its own money for the bridge through bond sales. But 1960 proved to be an unfavorable time for bringing a new bond issue to market, so he elected an approach that Harriman had rejected. He would arrange for the state to lend the money to the authority for a relatively short period. Under his plan, all the money would be repaid to the state from the proceeds of a Bridge Authority bond sale undertaken in a later, more hospitable market. The process proved to be more easily described than implemented, as it still had to be coordinated with the state's own year-by-year cash flows. The best the governor could do right away was to arrange for lending the authority about $4.5 million, to get started, and then to follow up the next year with the remaining approximately $18.5 million to pay for what was by then seen as about $23 million worth of construction.

The engineers may not have been entirely happy with that lingering uncertainty about financing, the governor's assurances notwithstanding, but they dealt with it creatively. Of the fifteen piers required to support the bridge, eight were very deep ones that had to be built well out in the river and had been designed as open concrete caissons like those used on Modjeski's Mid-Hudson Bridge at Poughkeepsie a little more than thirty years earlier. Of those eight, three had to be exceptionally deep, including one that was not expected to find satisfactory bearing until it reached a level about 165 feet below the surface of the water. Those three exceptionally deep piers would take longest to build, and they would be the most expensive, so Modjeski and Masters elected to isolate them as a separate contract and to advertise it for bids first. It was estimated that they would cost a total of about $4 million, an amount slightly less than the first advance the Bridge Authority had received from the state. In August 1960, when those bids were received, the lowest came from a joint venture of the Frederick Snare Company of New York and the Dravo Company of Pittsburgh, and it was for just under $4 million, a bulls-eye for the Bridge Authority. By October a contract had been signed with Snare-Dravo, and the first real step in the construction process had been taken.[14]

The Snare-Dravo joint venture was a powerful one. Dravo, which had been in business since 1891, was a giant in the marine construction business, capable of building almost anything related to rivers and harbors. It had a long history of building bridge foundations, docks, and tunnels, manufacturing marine equipment, and as a major player in the business of building barges and ships. During World War II Dravo had been essentially a shipbuilder and a major supplier of landing ships for tanks. Its base in Pittsburgh gave it ready access to a network of rivers, and signs of its presence were common wherever major work went on along any of the three rivers that converge there. Snare wasn't as big as Dravo, but it was a substantial and equally well-regarded heavy construction firm with almost sixty years of experience. Based in New York, it had done work in much of the eastern United States, in Cuba, and in the Caribbean islands, and it was of course no stranger to the Hudson, where it had twenty-seven years earlier been the general contractor for construction of the Rip Van Winkle Bridge.

The joint venture approach, popular in heavy construction, was apparently not one that had previously held much appeal for Dravo, which hadn't joined forces with another contractor since 1933, when the substructure for the San Francisco–Oakland Bay Bridge was considered too large for any single company. The Newburgh-Beacon probably struck Dravo's management as a propitious setting for another joint venture. They had done work for the Modjeski firm and were likely to have seen the size and nature of the Newburgh job as well suited to their own capabilities, although a little far from their normal base of operations. Snare certainly had the good reputation and substance to make it an attractive partner, and it had a good deal of equipment already positioned in the New York area. Even more important to Dravo, Snare had a long and well-established relationship with the labor unions that controlled construction labor on the Hudson, and it had a good following of men in the area.[15]

There was plenty of work to be done before Snare-Dravo could begin construction out in the river, although their field engineering crews would soon be establishing positions for the foundations. The real determinant for a starting date on the river was the fabrication of the bottom sections of the caissons, the ones with the integral cutting edges that would facilitate their being driven down into the silty bottom of the river. As in the case of the caisson-supported foundations for Modjeski's Mid-Hudson Bridge, these huge sections would be fabricated off-site, but this time even the off-site work would be done by one of the joint venture partners, not by an outside subcontractor. Since 1928 Dravo had been operating a big shipyard in Wilmington, Delaware, where it would build the cutting-edge sections for the Newburgh caissons, and by April 1961 the first of those big sections had been loaded onto a Dravo barge and was being towed up the Atlantic coast for delivery to Newburgh. Almost 90 feet long by 70 feet wide by about 20 feet high, with sharp teeth fabricated to make short work of the river's silt, it weighed 210 tons. One man who witnessed its arrival at the job site described it as being about the size of a six-room house.

By then, the state had advanced the rest of the required money to the Bridge Authority, and bids were taken for the remaining major elements of the work. To almost no one's surprise, Snare-Dravo submitted the lowest bid for the rest of the foundation work, which included the other five caisson-supported piers and a handful of others that would be built within cofferdams or would be founded on relatively shallow rock. This second contract was for about $5.2 million, bringing the total cost of the foundation work to about $8.5 million, almost half the cost of the bridge.

At about the same time, the two ranking American steel fabricators, Bethlehem Steel Company and the American Bridge division of U.S. Steel, submitted bids for the contract to furnish and erect the bridge's approximately 16,000 tons of steel. Each had previously won big steel contracts on Hudson River bridges, and this time the low bidder was Bethlehem Steel, at about $8.4 million. That bid, when added to the cost of the foundation work, meant that about three-quarters of the total estimated cost of the bridge had been confirmed by bids and that the project could be expected with confidence to be completed comfortably within its approximately $23 million budget. All that remained to be confirmed were contracts for roadway work and the administration building, and it was clear that there was more than enough money left in the budget to do those jobs.[16]

It was still 1961 when the contract was signed with Bethlehem Steel, but there was a good deal of work to be done before any steel would be erected. Snare-Dravo had to complete the piers, and Bethlehem itself had to order the steel from its mills, prepare and secure the engineers' approval for the complex shop details needed to fabricate it, do the required fabrication in its shops, ship the fabricated steel to a site near the bridge where the trusses could be assembled, and then deliver it all to the work site. Although a year wasn't a bit too long for those tasks, it was hoped that they might be completed sooner and would be well along by the beginning of 1962. In fact, that target wasn't missed by much. The first of the approximately 440-foot-long,

1,200-ton steel trusses was erected in June 1962, and by July 1963 Bethlehem's crews were getting close to setting the last trusses, the ones that would close the long gap across the shipping channel. By the end of the 1963 summer the topping out flag had been raised. During the fall, with the help of some good construction weather, the paving and painting that always come last were finished, and construction of the administration building was well advanced.

On the second of November 1963, with a light rain falling, Governor Rockefeller declared the approximately 1.4-mile-long bridge open, and he was joined by other dignitaries in congratulating everyone who had contributed to the project, assuring them that the bridge's very good prospects would justify all the hard work and years of waiting. The 30-foot-wide roadway was set up for two-way traffic, although it was understood that three lanes could be handled comfortably if and when necessary. The river below was almost entirely clear of barges and other construction-related vessels for the first time in almost three years. The 1,000-foot-long cantilevered structure across the main channel left almost 800 feet for shipping between the piers and 135 feet of clearance between high water and the bottom of the trusses.[17]

Almost everyone was pleased with the new bridge, but not everyone was in a festive mood. Three men had died in accidents on the work, and due respect was paid to them and their families by the opening day speakers. And the approximately seventy unemployed former ferry workers weren't happy either.

During the bridge's first few years, it's likely that most people felt pretty good about how it all seemed to be working out. No more ferry delays or cancellations, no more circuitous routes for drivers desperate to get to the other side of the river, and every indication that the work was well justified. But insiders were troubled by worrisome signs that the bridge was struggling to handle the large and increasing volume of traffic that Interstate 84 was bringing. In 1968, a year in which the bridge handled almost 5.5 million vehicles, the Bridge Authority tried to ameliorate the problem by increasing the number of toll booths, but that didn't change much. By 1969, when the traffic volume was well beyond the design capacity of a two-lane bridge, it was clear that the situation was becoming intolerable. The governor's 1960 decision to build a bridge with only two lanes had been based on an opinion that the need for greater capacity was twenty-five years off, but long waiting lines along the bridge approaches had by 1970 made it clear that his gamble had failed. Rockefeller brought Modjeski and Masters back to have a look at the traffic figures and to discuss approaches to solving the problem, and once he had their concurrence that a second bridge was going to be needed, he began a campaign for getting it built. It would be a long process, and the first step, of course, would be raising the money.[18]

Fortunately, the Federal Highway Administration had come a long way since 1960, when Rockefeller had decided to forgo federal aid in favor of advancing the money from New York's own treasury. Washington was still prepared to pay that once-elusive 90 percent of the cost, and federal tax collections were now steady enough to ensure relatively prompt release of the funds to the state. Rockefeller was more than ready to accept the government's terms, including its demand that the state agree to terminate toll collections as soon as soon as New York's own investment in the bridge had been recovered. In 1973 a document reflecting all those promises and constraints was signed by officials representing the Federal Highway Administration, the New York State Department of Transportation, and the New York State Bridge Authority, virtually assuring construction of a second bridge.[19]

The new plan produced by Modjeski and Masters described two steps that would ultimately provide a six-lane crossing between Newburgh and Beacon. As a first step, it proposed building a new bridge about 100 feet south of the first one and parallel to it. It would look almost exactly like the first bridge, except that its roadway would be wide enough for three generous traffic lanes instead of two and it would include an 8-foot-wide bicycle path as well. In the second step, the new three-lane south bridge would be used as a four-lane bridge for a few years while the original span was strengthened and widened and then reopened as a three-lane bridge. Once all that work was done, there would be six traffic lanes between Newburgh and Beacon.

It would be a while before the sometimes exasperating pace of government would allow approval of the plan and authorization for the engineers to begin the detailed design work required to implement it. To the casual observer, the proposed second bridge might have seemed a good deal more like the first one than it really was, so there may have been some impatience about its taking still another year to prepare documents for bidding. But the similarities weren't as pervasive as they seemed. For the engineers, it was almost like designing a brand new bridge.

New borings had to be made to verify that conditions under the new bridge would be just like those under the first one. The new bridge was to be more than half again as wide as the first one, an increase that would have engineering implications that went far beyond new dimensioning, and the foundations and structural frame to accommodate the revised loads and conditions would have to be designed as they were elements of an entirely new bridge. Even some of the design methodology would change: Unlike the original bridge, which was a riveted structure, the new bridge would be welded. The two bridges would end up looking almost identical, but they weren't.[20]

The players who would emerge to build the second bridge were different, too. A small contract for building a few land piers and the Newburgh abutment was awarded early in 1976 to a Westchester County contractor named Gardner Bishop, but it would be October before bids for the big caissons and other deep piers would be opened.

There would be no Dravo or Snare this time. In their place a joint venture of four other well-established heavy construction contractors would submit the winning bid. J. Rich Steers Inc. and Spearin, Preston and Burrows Inc., the firms that headed up the venture, were a couple of New York–based companies that had been doing marine construction since the early years of the century. The two other members of the joint venture were Yonkers Contracting Company, Inc., a heavy and highway builder that had done a good deal of work on the New York State Thruway and had an especially strong relationship with the labor unions that would control the workforce at Newburgh, and Buckley and Company, a heavy construction firm from Philadelphia that had recently completed a substantial marine construction contract in the Schuylkill River.[21]

And there would be no Bethlehem Steel Company on the second Newburgh bridge, either. A year after bids for the caisson and pier contract were opened, the American Bridge division of United States Steel Company pleasantly surprised the engineers and the Bridge Authority with a low bid of about $51 million for the superstructure contract, almost $17 million less than what had been budgeted. Two of the other five bids received for the superstructure contract were within about 3 percent of the lowest one, so there was implicit confirmation that the winning figure was not a fluke. The amount saved was welcome. Part of it was used to offset a small overrun in the substructure bidding, and there was still enough left over to protect against overruns that might surface in the work that remained to be purchased and to provide for any unanticipated expenses during construction. American Bridge was no newcomer to the Hudson Valley, of course, having built the steel superstructure for Modjeski's Mid-Hudson Bridge more than forty years earlier and having done the steel work on the Tappan Zee Bridge during the 1950s.[22]

There were still enough similarities between the new bridge and the old one to make the course and duration of its construction predictable. This time the substructure contractors elected to have the cutting-edge sections of the caissons fabricated locally, eliminating the long haul from Delaware that had probably slowed progress a bit on the first bridge. A fabricating facility was established on the Newburgh side of the river, and the sinking of the caissons began in 1977. It was well along in March 1978, when the worst fear of a substructure contractor was realized. One of the very deep caissons settled into a lopsided position about twenty degrees off the vertical as it was being driven down into the subsoil, and it remained stuck for about three months before a combination of ingenuity, brute force, and a rising tide wrenched it free.[23]

By the end of 1978 the pier work was approaching completion. When spring weather arrived early in 1979, an American Bridge workforce that had been mobilized during the winter was able to start erecting the approximately 24,000 tons of steel that had been fabricated in the company's mills in Pennsylvania, Indiana, and New York. Almost sixteen months later, the Bridge Company hoisted the last of its huge trusses from a barge in the river to complete the cantilever span across the shipping channel.[24]

What remained to be done on the second bridge after the steel work was in place was easy, compared with what had already been done, and it was all finished by October 1980. This time there was no waiting for the painter, because American Bridge had built the new structure using Corten, a recently developed "weathering" steel that promised to form a thin patina of rust to protect the underlying metal, obviating the time-consuming and costly process of applying a field coat of paint to the steel.

On an especially cold and blustery first day of November, in 1980, a modest crowd gathered to hear speeches by Senator Daniel P. Moynihan and other dignitaries and to cheer runners competing in a five-mile marathon that crossed the new bridge. Ribbons were cut at the Newburgh and Beacon entrances, a ceremonial parade of cars crossed, and the second span of the bridge was declared officially open by the chairman of the Bridge Authority.[25]

A little more than a month later, the original bridge was closed. The three lanes of the second bridge were temporarily converted to four, two of them eastbound and two of them westbound. It was a great improvement and broadly appreciated, but it wasn't yet the six-lane system that had been promised. That was still a few years off.

The task of strengthening and widening the original bridge to carry three lanes of traffic would be a complex one, in some ways more demanding than outsiders realized and certainly more expensive than some of the public had been allowed to believe. Local newspapers had been ominously silent on the matter of cost, but as recently as the middle of 1976 the *New York Times* had described the strengthening and widening as being a $20 million project.[26] That estimate may have originated with the lowest of a number of preliminary estimates made in the early 1970s, before the complexity of the work was fully understood and before the soaring inflation of the later 1970s had almost doubled the cost of similar construction. By 1981, when bids for work on the original bridge were about to be taken, Modjeski and Masters prepared detailed estimates for a variety of options for removing and replacing the bridge's deck, strengthening and widening its structural underpinnings, painting it, and generally bringing it up to current standards. Their estimates ranged from a little more than $40 million to almost $50 million, and in March 1981, seven bids were received for the work. Five of them fell within the predicted range, and the American Bridge Company's bid, at $43.5 million, was the lowest. By April a contract with American Bridge had been signed, and by June 1981 the work was underway.[27]

American Bridge engaged at least a couple of local subcontractors for the kind of highly specialized work that it was not prepared to do itself, such as road paving and lighting, but for the most part the work of strengthening and widening the first bridge was done by its own forces. It was immensely

FIGURE 15 Erecting the 500-foot-long, 1,800-ton truss assembly that would complete the superstructure framing for the second Newburgh-Beacon Bridge in August 1980. Photograph reproduced by permission of the American Bridge Company.

demanding work. Although there was enough of it to require all of the originally scheduled two and a half years to complete, the Bridge Company was almost finished a little sooner. Then the cold weather that often comes early in that part of the Hudson Valley took control, and completion of the concrete work for the new roadway and, most especially, the application of a recently developed Latex wearing surface for the roadway had to be delayed until the warmer weather of the next spring. By June 1984 everything was finished.

There having been two previous opening ceremonies for the Newburgh-Beacon Bridge (one in 1963 and the other in 1980), expectations for the third one were understandably low. The first opening had been addressed by the state's governor, the second by its senator, but this one would have to settle for speeches from members of the Bridge Authority and a few local political figures. Only about 150 people turned out to hear them, and it was just as well, as the weather turned rainy and windy. But the ribbons for the widened bridge were cut, and by the end of the day on 2 June 1984, six lanes were open and available for use on the two spans.[28]

FIGURE 16 The two Newburgh-Beacon bridges in late 1980, when the new span had been substantially completed but before the widening and strengthening of the original span had been started. Photograph courtesy of the New York State Bridge Authority.

Epilogue

IT DIDN'T TAKE LONG FOR THE TWO Newburgh-Beacon bridges to demonstrate their value. During their first full year of operation, almost 15 million vehicles crossed, an eminently satisfying average of more than 40,000 vehicles per day. Whatever excitement might have attended the opening ceremonies soon faded into memory, and it wasn't long before drivers on Interstate 84 began taking the pair of matching bridges pretty much for granted. Up and down the Hudson Valley, travelers had for years been accepting and eventually taking for granted their ability to cross the Hudson River at any one of an array of bridges and tunnels without having to drive more than a few miles to do it.[1]

The glory days of a great bridge are almost always limited to its earliest ones, when the larger-than-life product of years of work first appears above the water as a completed structure. For many of those involved—all the people who had done the planning, designing, and building—its completion had become an almost mythical goal that had finally been reached, and for some it was an experience that would never be repeated. A good sense of that unique connection between a builder and his bridge is wonderfully created in the character of the captured British colonel in Pierre Boulle's novel *The Bridge over the River Kwai*, who dies in an unsuccessful attempt to prevent his comrades from destroying the bridge he had just built for his Japanese captors. But the bridges of the Hudson were designed to perform, not just to gratify, and their stories all go well beyond their construction days.

The steel truss bridge that was built in 1909 to replace the original Waterford Bridge is an attractive structure designed by Albert Boller and Henry

Hodge. These distinguished New York engineers would later employ Howard Baird, who, after Boller and Hodge died, would go on to design the Bear Mountain Bridge and several others. Boller and Hodge's bridge at Waterford continues to function well today on its original, renovated piers, and its one-hundredth birthday was celebrated in 2009.[2]

A few miles downstream from Waterford, the Green Island Bridge, the first to carry a railroad across the Hudson, acquitted itself well during its relatively brief lifetime, but didn't prove as durable as its upstream neighbor. It managed to survive a series of damaging fires and was replaced several times before being undermined by a destructive flood in 1977. After that, it was replaced again, this time by a futuristic, metal-clad lift bridge that stands in the original location.[3]

The four other bridges that were subsequently built to cross the narrow reach of the river that forms Troy's western edge were unlikely to have been needed or built without the early growth and prosperity that had been brought to Troy by the Green Island Bridge and the railroad it carried. The first of those other four bridges was a wooden structure built at 112th Street in 1872 to connect Troy with Cohoes, and it was replaced in 1880 and again in 1923. The 1923 bridge was itself replaced in 1996 by the attractive beam-and-girder structure that stands today.[4]

The next Troy bridge came along soon after the Cohoes bridge and was completed in 1874. It connected Troy with Watervliet, the site of the great federal arsenal. Often called the Congress Street Bridge, it has been replaced three times, most recently in 1971.

During the twentieth century two other bridges were built at Troy. The Troy-Menands was the first. Built in 1932 and dedicated in July 1933 by Herbert H. Lehman, who had succeeded Franklin Roosevelt as governor of New York when Roosevelt became president, it had been designed as a lift bridge to ensure clearance for the high-masted vessels that still plied the river at that late date. By 1966 the vertical lift feature had outlived its usefulness, and its mechanisms and the operator's house were removed to save the cost of operating and maintaining them. It has continued ever since as a fixed bridge and provides a link to Interstate 787. In an interesting aside, all five of the engineers who managed the bridge's design and construction are considered sons of Troy, as all were graduates of Rensselaer Polytechnic, Troy's great engineering school. The most recent Troy crossing, the Collar City Bridge, sometimes called the Hoosick Street Bridge, was built in 1980 to carry Route 7 across the Hudson.[5]

South of Troy, in the Capital District that lies in and directly adjacent to Albany, the only survivor among the three railroad bridges that were built in the nineteenth century is the present version of the Livingston Avenue Bridge. The other two original railroad bridges and their turn-of-the-century replacements were demolished long ago in favor of structures designed to meet the modern, automobile-oriented demand that came with twentieth-century urban renewal. The original Livingston Avenue Bridge was reconstructed a couple of times, finally (around the turn of the century) as a steel railroad bridge, a formidable-looking black swing bridge that lies relatively close to the water and opens horizontally in a long arc to allow river shipping to pass through. It still carries freight and a few passenger trains across the river between Renssalaer and Albany, but it has been the subject of numerous studies that will doubtless one day lead to its replacement.

A short distance south of the Livingston Avenue, the imposing Parker Dunn Memorial Bridge now crosses the river about where the Albany-Greenburgh once stood. A large girder-and-stringer structure built in 1969 to replace its short-lived, more modest predecessor, the Parker Dunn was named for a World War I recipient of the Congressional Medal of Honor. It carries a periodically heavy volume of passenger and commercial vehicles into and out of downtown Albany. Two other Albany area structures of relatively recent vintage carry what is largely interstate highway traffic across the Hudson in the Capital District. One is the Berkshire Connector, an undistinguished truss bridge at Castleton, very close to the Castleton Cutoff that the New York Central built back in 1928 to enable its trains to bypass Albany. The other is the Patroon Island Bridge, a handsome deck truss span built just north of the city in 1968 to carry Interstate 90.

South of the Capital District, two bridges built around the middle of the twentieth century have remained essentially unchanged, and they continue to perform as expected. At the Rip Van Winkle, between Catskill and Hudson, more than five million vehicles crossed in 2007. Its toll booths were rebuilt in the early 1960s to accommodate a decision to collect tolls in only one direction, and the plaza was rebuilt to suit. A maintenance building was added during the 1990s, its architecture carefully conceived to preserve the Dutch Colonial style of the original administration building.

A few miles downstream, at Kingston, the bridge that connects Kingston with East Rhinebeck continues to be known by its geographically misleading name, the Kingston-Rhinecliff Bridge. The "Rhinecliff" in its name continues to honor the eastern terminus of the ferry the bridge displaced. Complicating the confusion about the bridge's name, in 2000 it was ceremonially renamed the George Clinton Kingston-Rhinecliff Bridge, after the man who served as governor of New York State from 1777 to 1795 and as vice president of the United States under Thomas Jefferson and James Madison. But old habits die hard, and almost everyone still knows the bridge as the Kingston-Rhinecliff. It has been carrying almost half again as many vehicles as its upstream neighbor between Catskill and Hudson, but the two bridges serve similar, mostly local constituencies. During the 1990s extensive repairs were made to its piers, and changes were made to its superstructure framing.[6]

The elegant Mid-Hudson Bridge, between Poughkeepsie and Highland, remains the graceful centerpiece of the Hudson Valley bridge system, and in 2007 it carried almost fourteen million vehicles. Increasing traffic has led to changes in its approach systems and its three toll lanes, which are regularly reconfigured to meet changes in demand. In 1993 the American Society of Civil Engineers designated the Mid-Hudson a New York State Civil Engineering Landmark, and in 1994 the State of New York formally renamed it the Franklin D. Roosevelt Mid-Hudson Bridge. But like its sister bridges on the Hudson, it continues to be called by its original name, the Mid-Hudson. It has a minor claim to fame that is probably independent of its beauty or its technological excellence. In 1985 two telephone call boxes were installed on the span, each with a sign offering help and encouragement to a person contemplating suicide and inviting him or her to use the phone to be connected with a mental health facility. Between 1985 and about 1996 fifty persons used the phone, and all but one were dissuaded from jumping.[7]

Less than a mile north of the Mid-Hudson, the once spectacular Poughkeepsie Railroad Bridge of 1889 lies like a derelict hulk, prostrate on its rusting supports, a continuing reminder of the decline of the country's railroads. Ever since sparks from one of the few trains still using it in 1974 ignited a fire on the bridge, it has been closed to railroad traffic and has been the focus of arguments about who should repair it, who would be willing to demolish it, and even about who owns it. After thirty years of such bickering, a local volunteer organization that calls itself Walkway Over the Hudson may have found a future for the old relic and has acquired ownership rights. Estimates for what Walkway wants to do start at $10 million, but its members have raised enough money for some restoration. During

2009 they made substantial progress toward converting the bridge to a linear park, an elegant green space almost two miles long and 200 feet above the Hudson, beckoning walkers, hikers, cyclists, and others to share a unique and exhilarating view of the river that even the drivers of motor vehicles have never been able to savor.[8]

Fifteen miles south of the old Poughkeepsie Railroad Bridge, the Newburgh-Beacon Bridge is thriving in its role as a vital link on Interstate 84. With six lanes, its traffic volume of more than 25 million vehicles during 2007 was among the highest on the river. But despite its clear success as a traffic conduit, opinions about its impact on the Hudson Valley are mixed. The city of Newburgh had been starting to show some economic distress even before the first of the twin bridges was built, but it had until then remained a hub of regional activity. Once the ferry was replaced by the first bridge in 1963, most travelers simply bypassed downtown Newburgh along Interstate 84, hardly noticing the grand old city on their way to or from other destinations. Newburgh has deteriorated badly since then, with most of its businesses moving out into commercial centers around the city, and although it would be unfair to attribute all its decline to the bridge, there's evidence that it was a factor. In 1997 the Newburgh-Beacon's name was formally changed to the Hamilton Fish Newburgh-Beacon Bridge, to honor one of the Hudson Valley's favorite sons. Four men named Hamilton Fish have represented the area in Congress and in other high places over the years, so it's not easy to determine which one is the honoree, but it is widely thought that the renaming of the bridge was intended as a tribute to all four. In fact, most people in the area still simply call the bridge the Newburgh-Beacon.

The local impact of the Bear Mountain Bridge, about eighteen miles farther downstream, has for the most part been more benign. As part of the Appalachian Trail and as the route of choice for entering and leaving the recreation areas of Bear Mountain and Harriman state parks, its contribution to the region's well-being has exceeded most expectations. On the other hand, a maxim that originated with politicians in New York City some years ago has given the bridge a little unanticipated and unwelcome attention. The maxim holds that anything a New York politician does after he crosses the Bear Mountain Bridge heading west shouldn't count against him, and it has been invoked, from time to time, to justify bad behavior. The bridge carried more than six million vehicles in 2007, and its restrictive load limits have inhibited the level of wear and tear normally imposed by heavier commercial vehicles, so its condition remains good. And it certainly seems to have laid to rest the early criticism that it was a blemish on the treasured landscape of the river. Most travelers now regard it as a welcome and graceful feature of the region.[9]

Closer to New York City, the Tappan Zee Bridge has been an effective if sometimes troubled bridge for years. Traffic increases from Westchester and New York and from unanticipated highway development in New Jersey have raised its average daily traffic to about 135,000 vehicles, a formidable burden for a bridge designed to handle only 100,000. Long lines and delays are common, and the bridge's unhappy patrons are numerous and outspoken, complaining about almost everything from the bridge's architecture to its safety and its toll rates. A number of renovations have been made over the years, and a movable barrier system has improved traffic flow by changing the number of lanes in each direction as conditions demand, but most of the complaints have been about conditions that are beyond economically feasible remediation. As the bridge reached and passed the fiftieth anniversary of its completion, the scope and estimated cost of maintaining it grew, and interest in the possibility of replacing it with a new bridge began to get some attention. More recently, concerns about the ability of the present bridge to survive an earthquake have been raised, further supporting an evolving case for a new bridge, and when proposals for a new intercounty bus system and a light rail system entered the public discourse, the case for building a new bridge became patently realistic. In 2008 the New York State Department of Transportation announced that it was preparing a plan to replace the bridge, estimating a total cost of about $16 billion, including about $6 billion for the new bridge alone and the balance for transportation systems and related work.[10]

Twenty miles south of the Tappan Zee, the George Washington is, at seventy-seven, almost as splendid a bridge as it was at sixteen, when the architect Le Corbusier described it as the most beautiful bridge in the world.[11] It survived the 1962 addition of a lower deck and trusses, which provided a much-needed increase in the number of traffic lanes it could carry but compromised the original breathtakingly slender north and south views of the bridge that had come to be treasured by almost anyone lucky enough to see them. In 2007 the George Washington carried more than 100 million vehicles.

There's reason to believe that the George Washington might be remembered even longer than the other bridges and tunnels of the Hudson. David

Aaron Rockland's engaging book about it cites and explores no fewer than a dozen published works of fiction and half a dozen poems in which the bridge figures prominently.[12]

The tunnels that cross the Hudson between New York and New Jersey haven't changed much since they were built, except for the improvement of climate controls and signage in the vehicle tunnels and the upgrading of rail systems in the railroad tunnels. The Lincoln and Holland tunnels continue to have problems with periodic peak loads, a result of increasing suburban commuter populations, and they now carry more than 200,000 vehicles per day.

McAdoo's rail tunnels, which now carry the PATH trains, are operating at full capacity. The September 2001 attack on the World Trade Center disabled one of them, and heroic action by the operators of the trains was responsible for rescuing two trainloads of passengers caught in the collapse of the buildings. The tunnels have been fully restored, but PATH is struggling to handle heavy passenger volume that has increased as commuters have turned away from automobile travel.

The history-making tunnels the Pennsylvania Railroad built back in the early days of the twentieth century began to become expensive luxuries soon after long-distance rail travel entered its decline, and by not much later than the middle of the century they were handling barely enough traffic to justify their existence. But it didn't take long for the growth of New Jersey's suburban population to change that, and by the 1970s those big tunnels had become part of the commuter rail system of New Jersey. In 2007 the capacity of the tunnels was being tested by a rail fleet that was carrying a daily average of about 150,000 passengers into and out of New York, and in June 2009 ground was broken for an additional commuter rail tunnel expected to be completed by 2017 and to cost $8.7 billion.[13]

NOTES

INTRODUCTION

1. James L. Stokesbury, *A Short History of the American Revolution* (New York: William Morrow, 1991), 82. A letter to the author from Professor Robert W. Venables of Cornell University, dated 23 June 2003, supports the (occasionally disputed) view that control of the Hudson was central to the strategies of both sides in the war.

2. Lincoln Diamant, *Chaining the Hudson: The Fight for the River in the American Revolution* (New York: Carol Publishing Group, 1989), 39–42.

3. Stokesbury, *Short History of the American Revolution*, 142–61.

4. Clare Brandt, *The Man in the Mirror: A Life of Benedict Arnold* (New York: Random House, 1994), 242–273.

1. WATERFORD, THE FIRST BRIDGE

1. James L. Stokesbury, *A Short History of the American Revolution* (New York: William Morrow, 1991), 249–57.

2. Descriptions and historical information about the North River sloop (sometimes called the Hudson River sloop) are based on material in *Hudson River Sloops*, a pamphlet privately published in 1970 by Hudson River Sloops Restorations Inc., available from that source at P.O. Box 25, Cold Spring, NY 10516.

3. The emigration of a small group of Nantucket whaling families to Hudson, NY, is described in an unpublished, undated pamphlet written by Charles S. Clark, a descendant of one of the families. It is entitled *The Emigration from Nantucket to Hudson, New York* and was made available to the author by Ms. S. Gardner of the Nantucket Historical Association, P.O. Box 1016, Nantucket, MA 02554. Details of the new settlement and the founding of the town of Hudson by the Nantucket "proprietors" (and how they fared afterward) are further explored by Margaret B. Schram in *Hudson's Merchants and Whalers: The Rise and Fall of a River Port, 1783–1850* (Hensonville, NY: Black Dome Press, 2004), 12–26.

4. Population growth in the Hudson Valley has been estimated from censuses that were taken every ten years by the federal government, beginning in 1790, and

privately published. Those data were republished in 1990 by the Norman Ross Publishing Company of New York as a series called *Decentennial Census of the United States,* and bound as individual volumes entitled *First Census of the United States, 1790; Second Census of the United States, 1800;* etc. Estimates used here are based on analysis of the populations of the ten counties in New York State that form the Hudson Valley, as they were calculated in the first five volumes of the *Decentennial Census.*

5. Daniel B. Klein and John Majewski, "Economy, Community and the Law: The Turnpike Movement in New York, 1797–1845," *Law and Society Review* 26, no. 3 (1992), 469–512.

6. A. J. Weise, *History of Lansingburgh* (Troy, NY: Edward Green, 1877), 19–20.

7. Information about the Granville Turnpike comes from a pamphlet called *Lansingburgh, New York: 1771–1971,* published by the Lansingburgh Historical Society but not dated. Additional information about the Granville Turnpike comes from an unpublished article written by Clarence E. Holden, the local historian for the Village of Whitehall, NY, from 1869 to 1927, and preserved in the collection of the Whitehall Historical Society. Information about the Whitehall Turnpike comes from S. E. Hammersley's *History of Waterford* (Waterford, NY: privately published, 1957) and from *Articles of Incorporation of the Waterford and Whitehall Turnpike Corporation.* The first is available in the collection of the Waterford Historical Museum and Cultural Center, 2 Museum Lane, Waterford, NY, and the other is preserved in the collection of the Historical Society of Whitehall, 12 Williams Street, Whitehall, NY. The story about the Rupert Turnpike comes from an article by Don Rittner, an upstate New York historian. It appears on the Internet at http://www.donritter.com/his.28.html as "Cross That Bridge When You Get to It."

8. The story of the Canajoharie bridge is described in two historical files at the Montgomery County (NY) Department of History and Archives. One is a report to the New York State Senate made by the Canajoharie and Palatine Bridge Company, dated 25 February 1847, and the other is an undated excerpt from the *Proceedings of the Meetings of the Town of Canajoharie.* That historical file of the Montgomery County Department of History and Archives contains an undated, unsigned article that provides a good overview of the collapse and reconstruction, together with a description of the roles played by Burr and his brother. Material about Theodore Burr's life is based mostly on an unpublished, undated biographical sketch written by Richard S. Allen of Round Lake, NY, and included in the Burr material maintained at the Waterford Museum, and on a similar biography written by Allen and published by the Litchfield Hills Association as "Theodore Burr—Torringford, Connecticut," *The Lure of the Litchfield Hills* 7, no. 4 (no date), 11–13. The article can be seen at the Oxford Memorial Library, 8 Fort Hill Park, Oxford, NY.

9. Information about the size of the pier bases is taken from Weise's *History of Lansingburgh,* 19. That the stone was laid without mortar was established from text that accompanied photographs of the bridge piers that appeared in the *Troy Times,* 20 August 1960, when renovations were being planned.

10. Details of the bridge construction were obtained mostly from Hammersley's *History of Waterford,* 114–16, and from a ten-page, extremely thorough, unpublished

analysis of the bridge's construction written by Professor Frank Griggs, a document that is also available at the Waterford Historical Museum. Professor Griggs, who retired from teaching engineering in 2002, is working on a book about the history of the Waterford-Lansingburgh Bridge. Additional details of the bridge's construction can be obtained by inspection of a scale model that was built in 1901 and is on display at the museum.

11. "Old Highway Bridge at Waterford, NY," *Engineering News*, 1 June 1889, 496, 497.

12. "An Ancient Bridge," *Troy Press*, 19 February 1895.

13. "Union Bridge at Waterford Falls Prey to Flames," *Troy Sunday Record*, 24 June 1984.

14. "Theodore Burr, Bridge Builder," a speech delivered by Vera H. Wagner, 5 April 1959, to the Pennsylvania Historical Society, Harrisburg, PA. Wagner was then president of the Theodore Burr Covered Bridge Society of Pennsylvania, and her speech was based mainly on material she had obtained from an article about Burr written by Hubertis M. Cummings and published in the *Quarterly Journal of the Pennsylvania Historical Society* 23, no. 4 (October 1956).

2. Steam and a Bridge at Troy

1. Asa Briggs, *The Power of Steam: An Illustrated History of the World's Steam Age* (Chicago: University of Chicago Press, 1982), 24–37, 50–59. The work of Savery, Newcomen, and Watts is well described in Briggs's book; readers who want a more exhaustive account of the early development of steam power would do well to see Robert Wallace's *The History of the Steam Engine: From the Second Century Before the Christian Era to the Time of the Great Exhibition* (London: John Cassell, 1852), 17–52.

2. "James Rumsey," a typescript dated 10 July 2000 in the collection of the Shepherdstown Public Library, Shepherdstown, WV. Rumsey was not born in the Potomac River town of Shepherdstown and actually spent much of his adult life in what is now Berkeley Springs, Maryland, about fifty miles away. But Rumsey launched his successful steam-driven vessel from Sheperdstown in December 1787, and so the place claims him as a native son.

3. Thompson Westcott, *The Life of John Fitch, Inventor of the Steamboat* (Philadelphia: J. B. Lippincott, 1857), passim.

4. Kirkpatrick Sale, *The Fire of His Genius: Robert Fulton and the American Dream* (New York: Free Press, 2001), 32, 38, 82, 83, 86–95, 120–37.

5. George Dangerfield, "Gibbons v. Ogden," *American Heritage* 14, no. 6 (October 1963).

6. Jim Shaughnessy, *The Delaware & Hudson: The History of an Important Railroad* (Berkeley, CA: Howell-North Books, 1967), 89–91. Readers interested in the broader historical context for establishing this first leg of what would ultimately become the New York Central Railroad can consult F. Daniel Larkin's *Pioneer American Railroads: The Mohawk and Hudson & the Saratoga and Schenectady* (Fleischmanns, NY: Purple Mountain Press, 1995), 13–25.

7. Rutherford Hayner, *Troy and Rensselaer County, New York* (New York: Lewis Historical Publishing Company, 1925), 577–88.

8. Richard C. Cote, "Isaac Damon's Northampton," 9 January 1970 (a typescript in the collection of the Forbes Library, Northampton, MA). Cote's essay is concerned mainly with Damon's architecture for buildings. For more information about Damon's life and career, see "Captain Isaac Damon," *Hampshire Gazette* (Northampton, MA), 6 September 1886.

9. *Connecticut Biographical Dictionary*, 2nd ed. (St. Clair Shores, MI: Somerset Publishers, 1994), s.v. "Town, Ithiel."

10. Ithiel Town's drawings, including details of his truss designs, are at the Beinecke Library, Yale University.

11. "The Green Island Bridge," an unsigned and undated typescript in the collection of the Rensselaer County Historical Society in Troy, NY. This detailed description of the bridge, excerpted from the typescript, has been attributed to nineteenth-century author Freeman Hunt (1804–1858), who wrote it after he crossed the bridge in 1836, the year after its completion.

12. Arthur James Weise, *The City of Troy* (Troy: Edward Green, 1886).

13. "Green Island Bridge Collapse Recalled on Anniversary," *The Times Record*, 15 March 1984.

3. Three Railroad Bridges at Albany

1. Thomas Phelan and P. Thomas Carroll, *Hudson Mohawk Gateway: An Illustrated History* (Sun Valley, CA: American Historical Press, 2001), 41–53.

2. *Remonstrance of the Citizens of West Troy, Against the Erection of a Bridge Across the Hudson at Albany* (Troy, NY: N. Tuttle, 1841), 1–8.

3. Alvin F. Harlow, *The Road of the Century: The Story of the New York Central* (New York: Creative Age Press, 1947), 42–83.

4. Ibid., 113–37.

5. Edward Hungerford, *Men and Iron: The History of the New York Central* (New York: Thomas Y. Crowell, 1938), 135–54.

6. Quoted in George Rogers Howell, ed., *Bi-Centennial History of Albany* (New York: W. W. Munsell & Co., 1886), 493.

7. William A. Breach, "Opening Argument Against the Hudson River Bridge Company, at Albany," presented 18 November 1856 to the Circuit Court of New York State, New York State Library, Albany. Beach's arguments were exhaustive and thoroughly documented, but not sufficiently compelling to induce the court to issue an injunction against construction of the bridge.

8. William H. Seward, *Argument of William H. Seward in the Albany Bridge Case, before the United States Circuit Court* (Albany: Weed, Parsons and Company, 1859), 1–33.

9. "The New Jersey Bridge Cases: Decisions of the United States Supreme Court," *New York Times*, 4 February 1862. The title of this citation is misleading. The original article had appeared in the *Newark Advertiser*, which wrote a headline that emphasized the New Jersey component of the story. The story was later copied, acknowledged, and published by the *New York Times*, this time emphasizing the New York bridge, but the original headline survived. In both publications, the story and outcome of the Albany case was included in the article.

10. "The Bridge at Albany," *New York Times,* 24 February 1866.

11. Joel Munsell, ed., *Collections on the History of Albany, from Its Discovery to the Present Time,* vol. 3 (Albany: J. Munsell, 1867), 287–90.

12. Ibid., 290–91.

13. Ibid., 292.

14. "The Bridge at Albany," *New York Times,* 23 February 1866.

15. "The New Bridge Bill," *Albany Evening Journal,* 29 April 1868.

16. "The New Bridge Bill," *Albany Argus,* 10 May 1869.

17. "The New Bridge Completed," *Albany Morning Express,* 29 December 1871.

18. Thomas R. Winpenny, *Without Fitting, Filing, or Chipping: An Illustrated History of the Phoenix Bridge Company* (Easton, PA: Canal History and Technology Press, 1996), 1–27. Phoenix Bridge would go on to a spectacular, but not wholly untarnished, role in bridge building after it finished its work at Albany. In 1907 a cantilevered steel bridge it was building near Quebec suffered a catastrophic failure, killing seventy-five workers, spilling 19,000 tons of steel into the St. Lawrence River, destroying the reputation of the eminent civil engineer who held ultimate responsibility for the design, and—despite the largely exonerating findings of an investigation—casting a dark shadow that would pretty much follow the company until its ultimate dissolution in 1984.

19. Winpenny, *Without Fitting, Filing, or Chipping,* 20–24.

20. John F. Alden, "Review of the Wrought Iron Girder Bridge across the Hudson River at Albany, NY" (senior thesis, Rensselaer Polytechnic Institute, 1872). Additional details of the bridge can be seen in Arthur G. Baker's senior thesis, "Wrought Iron Lattice Girder Bridge" (Rensselaer Polytechnic Institute, 1876).

21. "Death of Thomas Leighton," *Rochester Union and Advertiser,* 12 February 1886.

22. State of New York, represented in Senate and Assembly, 18 April 1872, *An Act Authorizing the Construction of a Bridge across the Hudson River at the City of Albany, and Incorporating the Albany and Greenbush Bridge Company.*

23. State of New York, represented in Assembly, 16 Apriı 1875, *Report of the Committee on Commerce and Navigation in Relation to the Investigation of the Albany and Greenbush Bridge Company.*

24. "The Bridge at Albany," *Albany Argus,* 25 January 1882.

4. The Last of the Railroad Bridges

1. *New York Times,* 28 April 1868.

2. *New York Times,* 14 March 1870. In this article, the names of some of Corning's associates on the Anthony's Nose project appear for the first time, including those of Andrew Carnegie and Generals Serrell and McClellan.

3. Edmund Platt, *The Eagle's History of Poughkeepsie: From the Earliest Settlements, 1683 to 1905* (Poughkeepsie: Platt and Platt, 1905), 201–39, 306.

4. Carleton Mabee, *Bridging the Hudson: The Poughkeepsie Railroad Bridge and Its Connecting Rail Lines* (Fleischmanns, NY: Purple Mountain Press, 2003), 9–13.

5. Henry Petroski, *Engineers of Dreams: Great Bridge Builders and the Spanning of America* (New York: Alfred A. Knopf, 1995), 22–66. Petroski's book, stunningly

thorough and well-written, explores the lives and work of some of the country's most important bridge builders. Because so many of the bridges he addresses were built across or near the Hudson, the book should be of great interest and value to anyone interested in the bridges of that region.

6. Ibid., 39–52.

7. Mabee, *Bridging the Hudson,* 15.

8. Harold C. Livesay, *Andrew Carnegie and the Rise of Big Business* (Glenview, IL: Scott Foresman & Co., 1975), 54–58.

9. Joseph Frazier Wall, *Andrew Carnegie* (Pittsburgh: University of Pittsburgh Press, 1970), 227–77.

10. "The Poughkeepsie Bridge and the Central New England and Western Railroad Line," *Poughkeepsie Daily Eagle* (souvenir edition), 1889. This article is an approximately 20,000-word history of the bridge that was included in the souvenir edition that was published in a single issue of the *Poughkeepsie Daily Eagle* in 1889. No more precise date of publication is shown. The full souvenir edition, including this article, is accessible in the collection of the Adriance Memorial Library in Poughkeepsie.

11. Ibid.

12. Ibid.

13. Lewis Beach, *Cornwall* (Newburgh: E. M. Ruttenber & Son, 1873), 164, 165.

14. "The Poughkeepsie Bridge: Description of the Proposed Structure as Called For by the Contract," *New York Times,* 14 February 1876. Readers knowledgeable about the history of American steel construction might (but should not) confuse this American Bridge Company of Chicago with the later, much larger, and more widely known American Bridge Company, the giant steel fabricating and erecting company established by U.S. Steel in Pittsburgh at the turn of the twentieth century to consolidate its vast steel fabricating and erecting holdings.

15. *Poughkeepsie Daily Eagle* (souvenir edition), 1889.

16. Ibid.

17. Ibid.

18. The history of Union Bridge and its five principals is compiled from a variety of sources, including "The Early Days of the Bridge Companies," an unpublished, undated memoir by Charles F. Kellogg, whose father was one of the original principals of a predecessor bridge construction company in Athens, PA. See also Thomas R. Winpenny, *Without Fitting, Filing, or Chipping: An Illustrated History of the Phoenix Bridge Company* (Easton, PA: Canal History and Technology Press, 1996), 15–27, and a report describing the groundbreaking for the Michigan Central Railway's cantilevered bridge at Niagara Falls in the *Niagara Falls Gazette,* 15 April 1883.

19. David B. Steinman, *Bridges and Their Builders* (New York; Dover Publications, 1957), 248, 249.

20. John F. O'Rourke, "The Construction of the Poughkeepsie Bridge," *Transactions of the American Society of Civil Engineers* 18 (June 1888), 200.

21. Thomas C. Clarke, "The Hudson River Bridge at Poughkeepsie," *Scientific American Supplement,* 19 May 1888.

22. *Poughkeepsie Daily Eagle* (souvenir edition), 1889.

23. Virtually all the descriptions of the construction process itself come from John F. O'Rourke's extensive and detailed "Construction of the Poughkeepsie Bridge."

24. Mabee, *Bridging the Hudson* 57–65.

25. "Castleton Cutoff of the New York Central Railroad," *Railway Review*, 14 July 1923, 41–43.

26. Letter to the author, 21 February 2007, from Richard Barrett, a scholar of Albany's railroad history who is a member of the board of directors of the New York Central System Historical Society and a member of the National Railway Historical Society. Barrett retired after thirty-nine years in government that included service as Deputy Commissioner of Public Works and as Commissioner of Parks and Recreation in Albany, NY.

27. "Castleton Cutoff," 49–52.

28. Author's interview with Richard Barrett, February 2007.

5. The Railroad Tunnels

1. Arthur G. Adams, *The Hudson through the Years*, 3rd ed. (New York: Fordham University Press, 1998), 123–35.

2. U.S. Bureau of the Census, Twelfth Census of the United States (1900), Census Reports, Volume 1 ("Population"), table 8, "Population of Incorporated Cities, Towns, Villages, and Boroughs . . . in 1900," http://www.census.gov/prod/www/abs/decennial/1800.htm.

3. "D. C. Haskin," *Engineering News*, Supplement (9 August 1900), 21.

4. "Down under the Hudson," *New York Times*, 17 July 1880.

5. "20 Men Buried Alive," *New York Times*, 22 July 1880.

6. "The Originator of the Hudson River Tunnel," *Engineering News*, 7 April 1904, 343.

7. William G. McAdoo, *Crowded Years: The Reminiscences of William G. McAdoo* (Boston: Houghton Mifflin, 1931), 67–78.

8. "C. M. Jacobs Dies," *New York Times*, 12 September 1919.

9. McAdoo, *Crowded Years*, 74, 75.

10. Henry Petroski, *Engineers of Dreams: Great Bridge Builders and the Spanning of America* (New York: Alfred A. Knopf, 1995), 122–30.

11. "Bridging the North River," *New York Times*, 5 January 1888.

12. Petroski, *Engineers of Dreams*, 130, 131.

13. Jill Jonnes, *Conquering Gotham, a Gilded Age Epic: The Construction of Penn Station and Its Tunnels* (New York: Viking, 2007), 40–54.

14. Cassatt was already well known, but he wasn't the only distinguished member of his family. His sister, Mary Cassatt, was one of the few members of the American Impressionist school of painters to be accepted by the leading French Impressionists, including Edgar Degas and Auguste Renoir. She was also the only woman known to be a member of either group.

15. Jonnes, *Conquering Gotham*, 57–59.

16. McAdoo, *Crowded Years*, 78–86.

17. Ibid., 87–98. McAdoo would prove to be a bold thinker in many ways during

his long life. When his interurban railroad was nearing completion in 1907, he began organizing a staff to operate the terminal. He hired women as ticket sellers and voluntarily paid them at the same rate as men working at the same level. It was a radical move for the times (more than ten years before women won the constitutional right to vote) and at first generated some resentment among his managers. But the decision proved to be a financial success when the women outperformed the men. McAdoo became a favorite of the growing women's suffrage movement, although he admits he had never paid any attention to the issue before he made that hiring decision.

18. Charles W. Raymond, "The New York Tunnel Extension of the Pennsylvania Railroad," *Transactions of the American Society of Civil Engineers,* paper no. 1150 (September 1910).

19. Charles M. Jacobs, "The New York Tunnel Extension of the Pennsylvania Railroad: The North River Division," *Transactions of the American Society of Civil Engineers,* paper no. 1151 (September 1910).

20. Jacobs, "New York Tunnel Extension."

21. "O'Rourke Will Build Pennsylvania Tunnel," *New York Times,* 12 March 1904.

22. Jacobs, "New York Tunnel Extension."

23. Ibid.

24. Ibid.

25. Ibid.

26. "H and M Gets Receiver," *New York Times,* 24 November 1964.

27. Brian J. Cudahy, *Rails under the Mighty Hudson: The Story of the Hudson Tubes, the Pennsy Tunnels and Manhattan Transfer,* 2nd ed. (New York: Fordham University Press, 2002), 57–65.

28. Joseph R. Daughen and Peter Binzen, *The Wreck of the Penn Central* (Boston: Little, Brown, 1971), passim.

6. The Bear Mountain Bridge

1. George Kennan, *E. H. Harriman: A Biography* (New York: Houghton Mifflin Company, 1922), 1–14.

2. Ibid., 15, 16.

3. Maury Klein, *The Life and Legend of E. H. Harriman* (Chapel Hill: University of North Carolina Press, 2000), 38–42.

4. Ibid., 44–47.

5. Persia Crawford Campbell, *Mary Williamson Harriman* (New York: Columbia University Press, 1960), 1–5.

6. Klein, *Harriman,* 49–60.

7. Ibid., 61–85.

8. Michelle P. Figliomeni, *E. H. Harriman at Arden Farms* (Arden, NY: Orange County Historical Society, 1997), 1–43.

9. Ibid., 45–96.

10. Robert O. Binnewies, *Palisades: 100,000 Acres in 100 Years* (New York: Fordham University Press, 2001), 45–96.

11. Ibid., 52–56.

12. John B. Rae, *The American Automobile: A Brief History* (Chicago: University of Chicago Press, 1965), 45–135.

13. "Frederick Tench, a Steel Engineer," *New York Times,* 28 October 1944.

14. Harriman to Tench, 12 January 1922, and Tench's response, 17 January 1922. Copies of both documents are held in the Arden House collection at the Orange County Historical Society, Arden, NY (hereafter OCHS).

15. Highlights of Howard Baird's career are given in *Who's Who in Engineering,* 3rd ed. (New York: Lewis Historical Publishing Company, 1931), but nothing is said there about his engineering education. In 1897, ten years into Baird's seventeen-year career at Phoenix Bridge Company, he was accepted into membership in the Engineers Club of Philadelphia, according to volume 14 of the *Proceedings of the Engineers Club of Philadelphia.* That recognition suggests that by 1897 he had in some way become educated in civil engineering. A check of all the schools within commuting distance of Phoenixville that were teaching civil engineering during the latter part of the nineteenth century failed to reveal evidence that Baird had studied at any of them. I assume that he had by 1897 learned at Phoenix Bridge all he needed to know to assume engineering responsibilities. His career after that, first as a structural engineer at Phoenix Bridge and then at the Brooklyn Rapid Transit Company, later at Boller and Hodge, and finally his acceptance for registration by New York State, as well as his ongoing work as an independent practitioner are seen as supporting that assumption.

16. Harriman to Tench, 12 June 1922, OCHS.

17. *Capitalization Statement of the Bear Mountain Hudson River Bridge Company.* This document announces and outlines the details of the sale of 7 percent Sinking Fund Bonds to generate $3 million, and the sale of 8 percent Income Bonds to generate another $1.5 million, as well as the distribution of 17,500 shares of common stock as a premium to the bond purchasers. The bond distribution is the subject of a letter from Bear Mountain Hudson River Bridge Company to W. A. Harriman & Co., Inc., 19 April 1923. Copies of these materials are held at OCHS.

18. *Agreement between Bear Mountain Hudson River Bridge Company and The Terry and Tench Company, Inc., 24 March 1923.* This document is preserved in the archives of the New York State Library, Albany.

19. "Towers of the First Suspension Bridge over Hudson River Completed," *Engineering News-Record,* 3 April 1924.

20. "Bear Mountain Suspension Bridge over Hudson River," *Engineering News-Record,* 24 December 1924. The length of time needed to spin the cables on the Brooklyn Bridge is taken from David McCullough's *The Great Bridge: The Epic Story of the Building of the Brooklyn Bridge* (New York: Simon and Schuster, 1972), 451.

21. Tench wasn't a complete stranger to bankruptcy, an all-too-frequent consequence of the hazards of contracting, and his earlier acquaintance with it adds an ironic note to his experience on the Bear Mountain job. His deceased father had been a contractor in Canada and had himself failed in 1888, leaving his creditors badly damaged. Twenty years later, young Tench had placed advertisements in a number of Canadian newspapers, inviting the creditors to submit evidence of his father's unpaid bills, and he followed by paying each of them in full, including

interest. This information taken from "Paying Old Debts," *Cass City (Michigan) Chronicle,* 12 August 1910.

22. "Building a Cliff Road to the Bear Mountain Bridge" *Engineering News-Record,* 6 November 1924.

23. J.V.W. Reynders to Roland Harriman, 8 December 1924, in the archives of the New York State Library, Albany.

24. "Unredeemed Ugliness in Bridge Construction," *Scientific American,* August 1924, 84.

25. Report of Bergen and Willvonseder, Certified Public Accountants, January 1940, as to the financial history and condition of the Bear Mountain Hudson River Bridge Company. Annual profits and losses of the bridge company are shown in the section called "Operations from April 1, 1925 to December 3, 1939." Document in the archives of the New York State Library, Albany.

26. Robert Moses to Governor Herbert H. Lehman, February 8, 1940, in the archives of the New York State Library, Albany.

7. The Holland and Lincoln Tunnels

1. "New-Jersey Bridge Report," *New York Times,* 2 September 1894.

2. Henry Petroski, *Engineers of Dreams: Great Bridge Builders and the Spanning of America* (New York: Alfred A. Knopf, 1995), 196–216.

3. "Propose $30 Million Dollar Suspension Bridge over the Hudson River with a New York Approach Near West 57th Street," *New York Times,* 22 December 1912.

4. "Highway Tunnels under Hudson Proposed by Engineers," *New York Times,* 5 May 1912.

5. "$42,000,000 Hudson Bridge," *New York Times,* 21 March 1913; and "Tunnels Not Bridge Favored to New Jersey," *New York Times,* 22 April 1913.

6. "Work Jointly for Tunnel," *New York Times,* 15 June 1918.

7. "Clifford Milburn Holland," *Dictionary of American Biography* as reproduced in the Biography Resource Center, Farmington Hills, MI, by The Gale Group, 2003.

8. D. A. Gasparini and Judith Wang, "Battery-Joralemon Street Tunnel," *Journal of Performance of Constructed Facilities* 20, no. 1 (February 2006), 92–107.

9. J. B. Walker, *Fifty Years of Rapid Transit, 1864–1917* (New York: Law Publishing Company, 1918). By 1907 the city's Board of Rapid Transit Commissioners, Holland's original employer, had done a spectacularly successful job of building the first subways. But once New Yorkers saw how wonderful the subways could be, they became angry that the board hadn't built more, so they saw to it that the board was dissolved. The governor responded by establishing the Pubic Service Commission (PSC) and giving it the power to supervise a range of public utilities, including New York's subways. Years later, management of subways would pass from the PSC to other commissions and agencies.

10. "Largest American Shield Tunnel Designed to Carry Vehicular Traffic under Hudson River," *Engineering News-Record,* 19 February 1920.

11. "Shield vs. Trench Method for Hudson Vehicle Tubes," *Engineering News-Record,* 5 May 1921.

12. Carl C. Gray and H. F. Hagen, *The Eighth Wonder* (Boston: B. F. Sturtevant Company, 1927), 45.

13. Ibid., 46–48.

14. Ibid., 49–60.

15. "Declares Goethals Tube Would Float," *New York Times,* 2 March 1920.

16. "Largest American Shield Tunnel," *Engineering News-Record,* 18 February 1920.

17. "New Hitch Delays Vehicular Tunnel," *New York Times,* 29 July 1921.

18. "Tube Bids Not Received," *New York Times,* 8 February 1922.

19. "$19,250,000 Lowest Jersey Tunnel Bid," *New York Times,* 16 February 1922.

20. "Bethlehem Gets Tunnel Contract," *New York Times,* 14 March 1922.

21. "Shields for New Tunnel," *New York Times,* 13 April 1922; and "Big Tunnel Caisson Slides from Ways," *New York Times,* 8 December 1922.

22. "Tunnel to Jersey Sets Speed Record," *New York Times,* 30 October 1923.

23. "Building the Hudson River Vehicle Tunnel," *Engineering News-Record,* 8 May 1924.

24. "Clifford Holland Dies after Breakdown," *New York Times,* 28 October 1924.

25. "Impressive Ceremonies in Two States Mark Opening of Holland Tunnel," *New York Times,* 13 November 1927.

26. "Seek Another Jersey Tube," *New York Times,* 2 March 1925; "Wants a New Hudson Tube," *New York Times,* 17 October 1927; "More River Tubes Needed," *New York Times,* 1 December 1927; "Two More Tunnels to Jersey Urged," *New York Times,* 23 January 1928.

27. Jameson W. Doig, *Empire on the Hudson: Entrepreneurial Vision and Political Power at the Port of New York Authority* (New York: Columbia University Press, 2001), 1–73, 165–71.

28. "Plan Second Highway Tunnel under Hudson River," *Engineering News-Record,* 13 March 1930.

29. "Hudson Tube Plan Held Up for a Year," *New York Times,* 5 March 1932.

30. "Hudson Tube Loan Is Accepted Here," *New York Times,* 1 September 1933.

31. "Engineers Named for 38th Street Tube," *New York Times,* 19 November 1933.

32. "Five Bid on Section of Midtown Tube," *New York Times,* 22 February 1934.

33. "Hudson Sandhogs Set Tunnel Mark," *New York Times,* 3 August 1935.

34. "Port Board Deal Frees PWA Funds," *New York Times,* 28 March 1935.

35. "Contract Let for Tube," *New York Times,* 5 February 1937.

36. "39th Street Tube Gets Name of Lincoln," *New York Times,* 17 April 1937.

37. "New Road under the Hudson," *Engineering News-Record,* 17 June 1937.

38. "Lincoln Tunnel Is Opened with Festive Ceremonies," *New York Times,* 22 December 1937.

39. "One Lincoln Tube Will Be Delayed," *New York Times,* 18 May 1938.

40. "New Lincoln Tube Will Open Today," *New York Times,* 1 February 1945.

41. "3d Lincoln Tube Is Opened; Big Test Due over Holiday," *New York Times,* 26 May 1957.

8. The George Washington Bridge

1. "For Hudson River Bridge above 125th Street," *New York Times*, 21 December 1923.

2. "Ferry Motor Traffic: 20% Increase This Season across Hudson River," *New York Times*, 2 July 1922. By the 1920s, Henry Ford was producing a car every fifteen minutes, more manufacturers were entering the industry every year, and the "installment plan" was making it easy to buy a car or a truck. The number of drivers seeking to cross the Hudson was much greater than (and the types of vehicles they were driving were much different from) anything anticipated by the ferry operators. The increasing volume of such traffic was making a multi-lane bridge look like a much better solution than another tunnel.

3. Jill Jonnes, *Conquering Gotham, a Gilded Age Epic: The Construction of Penn Station and Its Tunnels* (New York: Viking, 2007), 38–54.

4. Henry Petroski, *Engineers of Dreams: Great Bridge Builders and the Spanning of America* (New York: Alfred A. Knopf, 1995), 182–91.

5. Darl Rastorfer, *Six Bridges: The Legacy of Othmar H. Ammann* (New Haven: Yale University Press, 2000), 7. In addition to new and revealing information about Ammann, Rastorfer provides an exceptionally good (and singularly beautiful) collection of photographs showing his bridges during construction and after completion.

6. Ibid., 4–6.

7. Petroski, *Engineers of Dreams*, 223–27.

8. Ibid., 249–52.

9. Jameson W. Doig, *Empire on the Hudson: Entrepreneurial Vision and Political Power at the Port of New York Authority* (New York: Columbia University Press, 2001), 125–42.

10. Edward W. Stearns, "George Washington Bridge: Organization, Construction Procedure and Contract Provisions," *Transactions of the American Society of Civil Engineers* 97 (November 1933), 66–96.

11. Othmar H. Ammann, "George Washington Bridge: General Concept and Development of Design," *Transactions of the American Society of Civil Engineers* 97 (November 1933), 1–65.

12. Ibid.

13. Ibid.

14. "New Deck Begun on Bridge," *New York Times*, 2 June 1959.

15. Ammann, "George Washington Bridge."

16. Doig, *Empire on the Hudson*, 151.

17. "Says Hudson Bridge Will Deface Fort Washington Park," *New York Times*, 17 March 1925. Later, "Park Site Donated for Hudson Bridge," *New York Times*, 29 April 1927.

18. "3 Drown at Work on Hudson Caisson," *New York Times*, 24 December 1927. Useful insight into this tragedy, from an engineering perspective, can be found in "Construction of Substructure," *Transactions of the American Society of Civil Engineers* 97 (November 1933), 206, and in "Deep Cofferdam for Hudson River Bridge," *Engineering News-Record*, 16 August 1928.

19. "Reports Progress on Hudson Bridge," *New York Times*, 11 March 1928.

20. Ammann, "George Washington Bridge," 46–52.

21. Ammann, "George Washington Bridge." Also see *Inventing the Skyline: The Architecture of Cass Gilbert*, ed. Margaret Heilbrun (New York: Columbia University Press, 2000), 154–60. Gilbert wasn't the only architect to have something to say about the bridge. In 1947 the great French architect Le Corbusier visited America and described the George Washington as "the most beautiful bridge in the world." Le Corbusier, *When the Cathedrals Were White: A Journey to the Country of Timid People*, trans. Francis E. Hyslop Jr. (New York: McGraw-Hill, 1947), 75.

22. Petroski, *Engineers of Dreams*, 166–69.

23. Doig, *Empire on the Hudson*, 148–51. The story of the struggle against Governor Moore's effort to politicize the process is one of several to which Doig brings unique insights in this elegantly written book.

24. "Open Bids for Span across Hudson River," *New York Times*, 4 October 1927; also Ammann, "George Washington Bridge." The competition between eyebars and parallel wire cables, won by Roebling, was the main event, but it wasn't the only contest. The huge contract to fabricate and erect about 57,000 tons of structural steel for the bridge towers and deck was also at stake, and the competitors for that big prize were (not surprisingly) three giants: Bethlehem Steel Company, McClintic Marshall Company, and American Bridge Company. Bethlehem Steel was the company that U.S. Steel's Charles Schwab had brought to national prominence after he had split with J. P. Morgan early in the century. American Bridge was the company that Morgan had established by combining dozens of the country's largest fabricators into a single, powerful steel company. McClintic Marshall, for whom Ammann had worked as an engineer many years earlier, and which had recently provided steel for the nearby Bear Mountain Bridge, took the contract for the Hudson River bridge, but within a few years the steel picture would change. In 1931 Bethlehem Steel would acquire McClintic Marshall, creating a steel behemoth approximately equal in power to American Bridge. From then on, those two huge firms would call the shots on almost every big steel project in the country. That virtual monopoly continued until late in the twentieth century, when Japanese producers came along and changed everything.

25. "Work Is Speeded on Hudson Bridge," *New York Times*, 24 March 1929; also, "Erecting Towers and Floor Steel on the Hudson River Bridge," *Engineering News-Record*, 22 October 1931.

26. Ammann, "George Washington Bridge." See also: "How the Bridge Was Built," *New York Times*, 18 October 1931; "Spinning Four 36-Inch Cables for the Fort Lee Bridge," *Engineering News-Record*, 14 August 1930, 242–48; and "Last Wire of Span Spun over Hudson," *New York Times*, 8 August 1930.

27. "Hudson Span Named George Washington," *New York Times*, 24 April 1931. The directors who ratified the name may not have been aware (or may simply have not been troubled by the knowledge) that the Continental Army had lost the battle of Fort Washington. General Washington, having wisely decided to leave the fort before its final hours, had watched the surrender from a boat halfway across the Hudson, en route to Fort Lee.

28. "Two Governors Open Great Hudson Bridge as Throngs Look On," *New York Times*, 25 October 1931.

29. "Lower Deck of George Washington Bridge Is Opened," *New York Times,* 30 August 1962. The figure for the volume of traffic in 2007 comes from the Port Authority's website: www.panynj.gov.

9. The Mid-Hudson Bridge

1. The source for most of this story about the old Poughkeepsie Railroad Bridge, including details of the 1913 proposal to improve its pedestrian access and Gustav Lindenthal's dismissal of it as "impossible," is an unpublished document called "Historical Sketch of the Events Leading up to the Building of the Mid-Hudson Bridge at Poughkeepsie, New York," written by William R. Corwine and dated 9 October 1925. One copy was buried in the cornerstone of the Mid-Hudson Bridge. Another is held in the local history collection of the Adriance Memorial Library in Poughkeepsie.

2. "Report of Theodore D. Pratt" is part of the unpublished report of the executive committee of the Hudson Valley Bridge Association, undated but thought to have been distributed in February 1922. It is held in the local history collection of the Adriance Memorial Library, Poughkeepsie.

3. A summary of the Goethals report, which was included in the report of the executive committee of the Hudson Valley Bridge Association, is identified as "Report of George W. Goethals & Co." and is held by the Adriance Memorial Library in Poughkeepsie, in its local history collection.

4. "Memoirs of Deceased Members: George Washington Goethals," *Transactions of the American Society of Civil Engineers* 93 (1929). The ASCE memoir provides a good look at Goethals's career; additional details of his personal life can be found in "General Goethals Dies after Long Illness," *New York Times,* 22 January 1928.

5. Biographical material about Frederick Stuart Greene was obtained from the Preston Library at the Virginia Military Institute, Lexington, Virginia. Greene's papers are identified there as MS 0208.

6. "Memoirs of Deceased Members: Ralph Modjeski," *Transactions of the American Society of Civil Engineers* 106 (1941). Among the many engineers with whom Modjeski had been associated, sometimes as an employee, sometimes as a partner, and sometimes simply as a colleague, were three past presidents of the ASCE: George Morison, who was his first employer, starting about 1886; Alfred Noble, who became his partner in 1902; and George Smedley Webster, who became a fellow member of the board of engineers for the Delaware River Bridge project. Morison was a distinguished early builder of bridges himself, and he might have been the oddest of the lot. Young Modjeski remained in his employ for about seven years before moving on, and an article written years later by MIT Professor Elting Morison might provide some insight into his reasons for making the change. Professor Morison described his grand-uncle as an eccentric, often disagreeable, lifelong bachelor who lived alone in a fifty-seven-room house. See *American Heritage Magazine,* Fall 1989, available at http//:www.americanheritage.com/articles/magazine/it/1986/2/1986_2_34.shtml.

7. "Memoirs of Deceased Members: Daniel Edward Moran," *Transactions of the American Society of Civil Engineers,* 102 (1938). In 2008 the company founded by Moran was still in operation as Meuser Rutledge Consulting Engineers.

8. "Poughkeepsie's New Bridge," *Poughkeepsie Sunday Courier* (special edition), 11 October 1925.

9. Corwine, "Historical Sketch.."

10. "Smith Lays Cornerstone Today for Bridge," *New York Times,* 9 October 1925.

11. The story of the construction and sinking of the caissons, including details of the failure of the east caisson and of the long struggle to right it, has been assembled from a number of sources, starting with the report of the first launch ("Huge Bridge Caisson Launched," *New York Times,* 20 March 1927) and including the report that the east caisson was finally being righted ("Listed 19,000 Ton Caisson Righting Itself as Engineers Solve Hudson Bridge Problem," *New York Times,* 14 May 1928). The story of the display established on the dock comes from Rupert Sargent Holland, *Big Bridge* (Philadelphia: Macrae-Smith Company, 1934), 201, 202. The most comprehensive and reliable overview of the whole process, from beginning to end, is Glenn B. Woodruff, "An Overturned 19,000 Ton Caisson Successfully Salvaged," *Engineering News-Record,* 12 February 1932, and most of the story as told in this chapter derives from that article.

12. Moran's design to keep the top surface of the caissons depressed far below the water level was intended to reduce future interference with river navigation. The piers for the towers would be much smaller than the caissons that supported them, and they would occupy only a small fraction of the caissons' top surfaces. By keeping the caissons low, Moran made it possible for even a vessel with a deep hull to navigate very close to the face of the pier, which was clearly visible above the water, without trouble.

13. "Erection of 276-Ft. Towers for Mid-Hudson Suspension Bridge at Poughkeepsie," *Engineering News Record,* 2 October 1930.

14. Letter to the author from Michael Cegelski, senior vice president of American Bridge Company, describing the work done at Poughkeepsie by American Bridge, 12 August 2008.

15. "Mid-Hudson Bridge Paved in Alternate Slabs to Equalize Cable Loads," *Engineering News Record,* 2 October 1930. And the daily press didn't miss the rush to finish on time: "Rush New Hudson Span," *New York Times,* 1 August 1930.

16. "25,000 See Opening of Mid-Hudson Span," *New York Times,* 26 August 1930.

17. "Denies Cost of Bridge Has Been Excessive," *New York Times,* 18 July 1930.

18. "Ferries Plied the Hudson for 150 Years," *Poughkeepsie Journal,* 23 August 1980.

19. "4 Killed While Building Bridge," *Poughkeepsie Journal,* 23 August 1980.

20. Report of the 1999 sidewalk improvement comes from the website of the New York State Bridge Authority, http://nysba.net/bridgepages/MHB.

21. *Blakeslee Rollins Corporation, Appellant, v The State of New York, Respondent. Claim #19465, Court of Appeals of New York: 265 NY 567; 193 NE 323; 1934.*

22. *American Bridge Company, Inc., Respondent, v The State of New York, Appellant. Claim #22184, Supreme Court of New York, Appellate Division, Third Department: 245 A.D. 535; 283 MYS 577; 1935.*

10. The Rip Van Winkle Bridge

1. Details of the 1926 effort to launch the first plan for a bridge between Catskill and Hudson appear in a memorandum written on the letterhead of the State of New York, Executive Chamber, 6 May 1926. It was filed as an adjunct to Senate Bill Number 320, "An Act to provide for investigating the desirability of an additional bridge for vehicular traffic . . . between the village of Catskill and the city of Hudson. . . ." The memorandum, which is held in the local history collection of the Hudson Area Association Library, Hudson, NY, describes the legislation vetoed by Governor Smith and includes the note in which he directed the superintendent of public works to proceed with preliminary studies.

2. Raymond Beecher, "Rip's Hudson River Bridge," *Greene County Historical Journal* 9, no. 3 (Fall 1985), 21, 25–30.

3. National unemployment data appear in United States Bureau of the Census, *Historical Statistics of the United States, Colonial Times to 1970, Bicentennial Edition, 1976, Part 1*, 125–36. Figures for unemployment in Greene and Columbia counties come from a special unemployment census that was included in the *1930 Decennial Census*, 724.

4. Thomas E. Rinaldi and Robert J. Yasinac, *Hudson Valley Ruins: Forgotten Landmarks of an American Landscape* (Hanover, NH: University Press of New England, 2006), 52–59.

5. Richard P. O'Connor, "A History of Brickmaking in the Hudson Valley" (Ph.D. diss., University of Pennsylvania, 1987), 1–6, 311–26. Additional relevant information about the decline of brickmaking in the Hudson Valley is given in George V. Hutton's *The Great Hudson River Brick Industry* (Fleischmanns, NY: Purple Mountain Press, 2003), 132–65.

6. United States Bureau of the Census, *Historical Statistics of the United States, Colonial Times to 1970*. Bicentennial Edition, 1976, Part 2. 716.

7. Biographical material about Glenn Woodruff is provided in a memoir written by Thornton Corwin for *Transactions of the American Society of Civil Engineers* (1974), 580. Biographical information about Frederick Stuart Greene appears in Chapter 9, note 5, above, but a few further insights into the persona of that interesting and complex engineer are provided in an appreciation by Jessica Amanda Salmonson, "Frederick Stuart Greene (1870–1929), Horror's Discarded Genius" (http://www.violetbooks.com/greene.html).

8. Glenn Woodruff, "An Overturned 19,000 Ton Caisson Successfully Salvaged," *Engineering News-Record*, 12 February 1932.

9. A print of Woodruff's drawing, dated 30 July 1930, is held in the local history collection of the Hudson Area Association Library, Hudson, NY.

10. A. B. Greenleaf, "The Rip Van Winkle Bridge," *Roads and Streets* 68 (November 1935), 350–52.

11. Henry Petroski, *Engineers of Dreams: Great Bridge Builders and the Spanning of America* (New York: Alfred A Knopf, 1995), 101–11.

12. "World's Longest Bridge Spans," a tabulation by Jackson Durkee, PE. Dated 14 May 1999, it was prepared for the National Steel Bridge Alliance, a division of the American Institute for Steel Construction, One East Wacker Drive, Chicago.

13. Jesse H. Jones, *Fifty Billion Dollars: My Thirteen Years with the RFC, 1932–1945* (New York: Macmillan, 1951), 169, 170.

14. Beecher, "Rip's Hudson River Bridge."

15. Petroski, *Engineers of Dreams*, 303–7.

16. "NY Firm Is Low Bidder for Bridge Here," *Hudson Daily Star*, 2 June 1933.

17. Much of the information about Snare's history comes from interviews with Ted Seeley of Mt. Pleasant, SC, and from Chip Triest of Alexandria, LA. Seeley is a former employee of the Snare Corporation and is the grandson of the man who succeeded Frederick Snare as president. Triest is the grandson of Gustav Triest, Snare's former partner. Additional biographical material about both Frederick Snare and Gustav Triest was obtained from obituaries that appeared in the *New York Times* on 23 September 1946 and 22 September 1946, respectively. Both men died on the same day, 21 September 1946, Snare in Havana, Cuba, and Triest in Mineola, NY.

18. "Modern Construction Practices on Latest Hudson River Bridge," *Engineering News-Record*, 7 November 1935, 640–42.

19. Greenleaf, "The Rip Van Winkle Bridge," 350–52.

20. Some description of the steel-erecting process is given in *Roads and Streets* (November 1935), 350, 351. Additional information about it is given in the *Greene County Historical Journal* 9, no. 3 (Fall 1985), 26, 27.

21. "Open New Bridge across Hudson River," *Hudson Daily Star*, 3 July 1935.

22. Traffic volumes for the Mid-Hudson and Rip Van Winkle bridges are based on data developed by D. B. Steinman, Consulting Engineers, as part of that firm's October 1948 design proposal to the New York State Bridge Authority for the Kingston-Rhinecliff bridge. The proposal was submitted to the authority in October 1948 and is held in its archives at its headquarters in Highland, NY.

11. The Tappan Zee Bridge

1. "After 25 Years, Fred Horn's Prophecy Becomes a Reality," *Rockland Journal-News*, 14 December 1955.

2. "Nyack Bridge Plan Dropped by Board," *New York Times*, 20 November 1936.

3. Robert A. Caro, *The Power Broker: Robert Moses and the Fall of New York* (New York: Vintage Books, 1975). Caro's biography effectively explores the relationship between Madigan and Moses and goes on to examine in intimate and informative detail the world of engineering and construction in Depression-era New York, when many of the engineers and builders of later bridges were honing their skills. Also see "Madigan Leaves Engineering Post," *New York Times*, 1 August 1967, for a view of how Jack Madigan's career later played out.

4. "Emil Praeger, Ranking American Engineer," *Rockland Journal-News*, 14 December 1955. For information about some of the work Praeger would have been doing before he turned to bridges, see the biographical note on Bertram Goodhue in the finding aid to the Bertram Grosvenor Goodhue Architectural Drawings and Papers, Avery Architectural and Fine Arts Library, Columbia University.

5. Alfred Stanford, *Force Mulberry: The Planning and Installation of the Artificial Harbor off U.S. Normandy Beaches in World War II* (New York: William Morrow,

1975), passim. Much of the publicity later associated with Praeger, who retired from the navy as a four-stripe captain at the end of the war, describes his work on the Normandy program in language that makes his role look like a very senior, policy-making one. Praeger's own statements on that subject never implied such a role, and Commander Stanford's book never mentions him. What seems most likely is that Praeger had an important but not very senior role.

6. "New $5,000,000 Pier, First of Its Type, Is Planned by City," *New York Times*, 2 January 1950.

7. "Tappan Zee Span for Thruway Seen," *New York Times*, 27 April 1950.

8. The nature of the opposition is described and the names of a few of the opponents are given in "Fight on Tappan Zee Span," *New York Times*, 7 May 1950. "Port Body Gives In on Thruway Span," *New York Times*, 12 May 1950, tells how Governor Dewey won the argument with Governor Driscoll.

9. "Thruway's Links on River Specified," *New York Times*, 10 December 1950.

10. "Hudson River Bridge for Thruway Voted," *New York Times*, 15 March 1951.

11. "Army Engineers Approve Thruway Bridge Petition, Authority to Start Work," *Rockland Journal-News*, 2 May 1951.

12. "Nyack Waterfront Base for Bridge Construction," *Rockland Journal-News*, 4 June 1952.

13. Emil Praeger, "Foundation Problems, General Design and Structural Features, Nyack-Tarrytown Bridge," *Municipal Engineers Journal* 41, no. 3 (1955), 137–43.

14. Merritt-Chapman & Scott, a company once known as "The Black Horse of the Sea," had begun to add marine construction work to its marine salvage business during the early part of the twentieth century. In addition to such high-profile salvage jobs as the righting of the fire-damaged French liner *Normandie* during the early days of World War II, it had begun to do major bridge foundation work in much of the world. After completion of the Tappan Zee Bridge, it would work on the Kingston-Rhinecliff Bridge farther north on the Hudson and on many other major jobs. Later in the 1950s it would build the foundations for David Steinman's Mackinac Bridge in Michigan; still later it would work on the Throg's Neck Bridge in New York. But during the early 1970s the effects of management's efforts to expand into other businesses would prove too much, and the company became financially overextended and was liquidated.

15. "Building Big Buoyant Boxes for Bridge Substructure," *Engineering News-Record*, 9 July 1953.

16. "Nyack Bridge Plan Is Revised; No Bids," *New York Times*, 16 December 1952.

17. "Low Bid Received on Thruway Span," *New York Times*, 17 April 1953.

18. "First 250 Foot Steel Segment is Floated onto Thruway Bridge Supports Near Tarrytown," *New York Times*, 21 July 1954. Additional details of the contractor's ingenious use of the tides are given in "Tides in Hudson Played Star Role in Tappan Zee Bridge Building," *Rockland Journal-News*, 14 December 1955.

19. "Thruway Bridge Gets Falsework," *New York Times*, 24 October 1954.

20. "Tappan Zee Bridge, Suffern to Yonkers Thruway Opened," *Rockland Journal-News*, 15 December 1955.

21. "Tappan Zee Is Official; Governor Signs Bill Naming the Tappan Zee Bridge," *New York Times*, 29 February 1956.

12. The Kingston-Rhinecliff Bridge

1. Glendon Moffett, *The Old Skillypot and Other Ferryboats of Rondout, Kingston and Rhinecliff* (Fleischmanns, NY: Purple Mountain Press, 1997), 59–72. Moffett provides a good account of the ferry's history, but readers who want to know more can find a delightful account of the history of Rhinecliff itself, including its relationship with the ferry, in Cynthia Owen Philip, "Downtown and Upstreet: The Saga of Rhinecliff," in the Summer 2005 issue of *About Town*, published quarterly in Red Hook, NY.

2. Moffett, *Old Skillypot*, 88–93.

3. "Notes How Some Scoffed at Plan," *Kingston Daily Freeman*, 11 May 1957.

4. Material about Steinman's life and career is based mostly on William Ratigan, *Highways over Broad Waters: Life and Times of David B. Steinman, Bridgebuilder* (Grand Rapids, MI: William B. Eerdmans Publishing Company, 1959), 11–106, an informative but brazenly hagiographic biography. The feud with Ammann is well known, but Ratigan gives it only superficial attention without much explanation. It is explored in more depth in Dale Rastorfer's elegant (and beautifully illustrated) *Six Bridges: The Legacy of Othmar H. Ammann* (New Haven: Yale University Press, 2000), 8–11.

5. Ratigan, *Highways over Broad Waters*, 146–241. Steinman and Holton Robinson, who was a generation older than Steinman, maintained their unorthodox partnership until 1945, when Robinson died. They sometimes did jobs together, sometimes separately, and never had a written contract. Robinson had designed or consulted on the design of many of the country's most important bridges and was the inventor of a number of techniques and devices for erecting the complex components required for suspension bridges. Steinman regarded him as a master engineer and defers to him often in descriptions of his own work in *Bridges and Their Builders*, written with Sara Ruth Watson (1941; reprint, New York: Dover Publications, 1957), 301–4, 333–40, 388.

6. D. B. Steinman, Consulting Engineer, "Proposed Kingston-Rhinecliff Bridge over the Hudson River," a presentation to the New York State Bridge Authority, 1 October 1948. Typescript available in the collection of the Kingston Public library, Kingston, NY.

7. "Loughran Reports Bridge Location Is Not Yet Settled," *Kingston Daily Freeman*, 23 December 1949.

8. "Span to Be Four Miles North of Rondout," *Kingston Daily Freeman*, 4 December 1950.

9. Emerson W. Pugh, *Building IBM: Shaping an Industry and Its Technology* (Cambridge, MA: MIT Press, 1995), 145–82. Also, for a few details about ENIAC (the University of Pennsylvania's computer), see http://lecture.eingang.org/eniac.html.

10. "IBM Moves to New Kingston Plant," *The Kingston IBM News*, 25 November 1955.

11. Ratigan, *Highways over Broad Waters*, 277–313.

12. W. E. Joyce and C. H. Gronquist, "Design of the Kingston Bridge over the Hudson River," *Proceedings of the American Society of Civil Engineers* 81 (May 1955).

13. "Bridge Piers Bid $4,495,477," *Kingston Daily Freeman*, 29 April 1954.

14. "Bridge Contract Is Let," *Kingston Daily Freeman*, 5 August 1954. The other two bidders on the big superstructure contract were American Bridge Company (a division of U.S. Steel) and Bethlehem Steel Company, two giant fabricators that had both (like Harris) done big steel jobs on Hudson River bridges before.

15. "Erection Speed Governs Newest Hudson Bridge," *Engineering News-Record*, 9 August 1956.

16. Edmund Wilson, *Apologies to the Iroquois* (New York: Farrar, Straus and Cudahy, 1960), 3–36. Wilson's book collects articles he had published in *The New Yorker* about the Iroquois, but the chapter that deals specifically with the Kahwanake, a Mohawk community within the Iroquois Nation, was an exception. Written by Joseph Mitchell, it had appeared in *The New Yorker* with the title "The Mohawks in High Steel." Wilson apparently felt it was so important that he included the entire article in his book and gave Mitchell full credit for it.

17. "Bridge Opens Here over Hudson," *Kingston Daily Freeman*, 2 February 1957.

18. "Wickes Hopes Bridge Is Strong Area Bond," *Kingston Daily Freeman*, 11 May 1957.

13. The Newburgh-Beacon Bridges

1. "Hudson River Bridge Plan Old Not New," *Evening News* (Newburgh), 7 July 1959.

2. The competition with Poughkeepsie is well described in Carleton Mabee, *Bridging the Hudson: The Poughkeepsie Railroad Bridge and Its Connecting Rail Lines* (Fleischmanns, NY: Purple Mountain Press, 2001), 21, 22. Legislation barring bridge construction on the Hudson until completion of the bridge at Kingston (later repealed) is described in "Hudson River Bridge Plan Old Not New," *Evening News* (Newburgh), 7 July 1959.

3. The abundance of articles about the history of the Newburgh-Beacon ferry attests to its prominence in the public's thinking. One of the best was an article included in a pamphlet honoring the 250th anniversary of the City of Newburgh, 5–12 July 1959. Entitled "The Newburgh-Beacon Ferry," it appeared only a few years after the ferry had been taken over by the state and was written by Pauline Ramsdell Odell, whose family had owned the ferry for years. The pamphlet, *250th Celebration Newburgh, New York 1709–1959*, can be seen in the Local History Room of the Newburgh Free Library. Another informative piece about the ferry appeared many years later: "Ferry Tale Ended Twenty Years Ago," *Evening News* (Newburgh), 4 December 1983.

4. "As We See It," *Newburgh News*, 26 December 1956.

5. New York State Department of Public Works, *Report on the Proposed Newburgh-Beacon Bridge* (Albany, 1952). The report can be seen in the archives of the New York State Bridge Authority, Highland, NY.

6. *The Planners*, an overview of the bridge planning process written by Maurice D. Herbert, executive editor of the *Newburgh News*, was included in a brochure prepared by the bridge celebration committee and distributed at the bridge dedication, 2 November 1963. A copy is available in the Local History Room of the Newburgh Free Library.

7. An undated, unpublished manuscript arranged and written by Margaret W. Masters, "Memoir of Frank M. Masters, and Historical Sketch of Modjeski and Masters," is held in the archives of Modjeski and Masters, Consulting Engineers, in Harrisburg, PA. It was made available to the author by the Modjeski and Masters office in Harrisburg.

8. "Ferry Plan Proposed," *New York Times*, 31 August 1955.

9. U.S. Department of Transportation, Federal Highway Administration, *America's Highways, 1776–1976: A History of the Federal-Aid Program* (Washington, DC: Government Printing Office, 1976), 254–56.

10. Modjeski and Masters, *Report to the New York State Bridge Authority: Location and Design, Newburgh-Beacon Bridge. 19 November 1956*. The report can be seen at the New York State Library, Albany.

11. "Hudson River Bridge Plan Old Not New," *Evening News* (Newburgh), 7 July 1959.

12. Ibid.

13. "Dedication Will Mark Fulfillment of 13-Year Dream of a River Span," *Poughkeepsie Journal*, 29 October 1963. In its review of the history leading to construction of the two-lane bridge, the *Journal* article explores Rockefeller's decision and identifies some of the opposition he faced.

14. "Bridge Work to Start," *New York Times*, 6 October 1960.

15. Most of the information about Snare's history comes from interviews conducted during November and December 2008 with Ted Seeley, a former engineer in the Snare Corporation and the son of a former president of the company. Information about Dravo comes from *A Company of Uncommon Enterprise: The Story of Dravo Corporation, 1891–1966, the First 75 Years*, a privately published history that can be seen at the Pittsburgh Public Library.

16. "Newburgh-Beacon Bridge Nears Reality," *Evening News* (Newburgh), 15 October 1961.

17. "Ceremony Opens Newburgh Span," *New York Times*, 3 November 1963.

18. "Six Lane Bridge Endorsed between Newburgh, Beacon," *Poughkeepsie Journal*, 6 October 1972.

19. "U.S. Will Pay 90% of Cost of New Bridge at Newburgh," *New York Times*, 31 January 1973. The agreement to abandon toll collection at a later date was a condition of the funding in 1973, but fifteen years later a new agreement relaxed that requirement. Tolls are still being collected on the Newburgh–Beacon Bridge in 2009.

20. "Hudson Bridge Construction to Start in June, a Year Behind Plan," *New York Times*, 24 May 1976.

21. "N-B Bridge Low Bid at $36 Million," *Evening News* (Newburgh), 30 October 1976. The article reports the details of the bidding. Information about the histories

of the four joint-venture companies was obtained from publicity literature provided by the companies.

22. "Structure Bid $51 Million," *Evening News* (Newburgh), 7 October 1977.

23. "Bridge Caisson Still Stuck," *Evening News* (Newburgh), 10 March 1978. For comparison with a similar event on the Mid-Hudson Bridge project, where the delay was about a year, see Chapter 9.

24. "Newburgh-Beacon Span Finished," *Times-Herald Record* (Middletown), 2 August 1980.

25. "Second Span Opened," *Evening News* (Newburgh), 2 November 1980.

26. "Hudson River Construction to Start in June, A Year Behind Plan," *New York Times*, 24 May 1976. What enhanced the credibility of the incorrect information about the estimated cost of the widening was the fact that all the other cost information given in the article was correct.

27. "Overhaul of Old N-B Span Begins," *Evening News* (Newburgh), 26 June 1981.

28. "Ceremony Marks Reopening of Bridge," *Evening News* (Newburgh), 3 June 1984.

Epilogue

1. Crossing data for the Newburgh-Beacon bridges are shown by the New York State Bridge Authority on its website: http://nysba.state.ny.us.

2. Information about Boller and Hodge, about Howard Baird, and about Baird's later work on the Bear Mountain Bridge is given in Chapter 6. Details of the fire that brought down the original Waterford Bridge in 1909 can be found in "Union Bridge at Waterford Falls Prey to Flames," *The Record* (Troy, NY), 10 July 1909.

3. "Green Island Bridge Collapse Recalled on Anniversary," *The Times Record* (Troy, NY), 15 March 1984.

4. "Bridge Part of Area's History," *The Record* (Troy, NY), 13 November 1996.

5. Information about the twentieth-century bridges in the Troy area that to one extent or another can be seen as having been spawned by the Green Island Bridge is taken from "The Bridges of Troy," *The Record* (Troy, NY), 13 November 1996, and from *Troy Community Newsletter*, an undated newsletter that can be seen on the Internet at http://www.uncle-sams-home.com.

6. The website of the New York State Bridge Authority, at http://nysba.state.ny.us, is the source of most of the updated material about the Rip Van Winkle Bridge and the Kingston-Rhinecliff Bridge.

7. "Calls Stop Many a Jump off Mid-Hudson Bridge," *New York Times*, 29 February 1996.

8. "Landmarks: Rusty Bridge, Great Views and Soon, a Walkway?" *New York Times*, 21 January 2007.

9. "NYC: Sad but True, Missing the Old Days in Albany," *New York Times*, 25 March 2007.

10. "State to Replace, Not Rebuild, Tappan Zee Bridge," *New York Times*, 28 September 2008.

11. Le Corbusier, *When the Cathedrals Were White: A Journey to the Country of Timid People,* trans. Francis E. Hyslop Jr. (New York: McGraw-Hill, 1947).

12. David Aaron Rockland, *The George Washington Bridge: Poetry in Steel* (New Brunswick: Rutgers University Press, 2008).

13. "Transit Chief's To-Do List: No. 1, New Hudson Tunnel," *New York Times,* 6 May 2007.

INDEX

ABOUT THE AUTHOR

Donald E. Wolf, P.E., graduated from Cornell University and worked in engineering for about forty years before switching to writing about it. His first book, *Big Dams and Other Dreams* (1996) was about the joint venture contractors who built Hoover Dam and other big structures; and his second, *Turner's First Century* (2002) was the centennial history of the great Turner Construction Company. *Crossing the Hudson* is his third book.